Solutions Manual for

Techniques of Problem Solving

Solutions Manual for
Techniques of Problem Solving

Luis Fernández and Haedeh Gooransarab

with assistance from Steven G. Krantz

American Mathematical Society

ISBN 0-8218-0628-9

TABLE OF CONTENTS

Preface

This manual contains the solutions to most of the exercises in the book **Techniques of Problem Solving** by Steven G. Krantz, hereinafter referred to as "the text."

It is essential that this manual be used only as a reference, and never as a way to learn how to solve the exercises. It is strongly encouraged never to look up the solution of any exercise before attempting to solve it. The 'attempt time' will always be as rewarding to the student—or maybe more—as solving the exercise itself.

Notation and references as well as the results used to solve the problems are taken directly from the text.

We would like to express our gratitude to Nicola Arcozzi, Steven Krantz and Vladimir Maşek for their valuable comments on several of the exercises.

Luis Fernández and Haedeh Gooransarab.

St. Louis, April 1st 1996.

Chapter 1

Basic Concepts

1.1. Let us use induction. We will prove something that at first seems stronger: $\left(\sqrt{2}-1\right)^k = \sqrt{N_k} - \sqrt{N_k - 1}$ for N_k a positive integer satisfying $\sqrt{2}\sqrt{N_k}\sqrt{N_k - 1} \in \mathbf{Z}$. We added this last condition because it helps in the induction argument.

The statement is true for $k = 1$: picking $N_1 = 2$, we have

$$\left(\sqrt{2}-1\right) = \sqrt{2} - \sqrt{2-1},$$

and

$$\sqrt{2}\sqrt{1}\sqrt{2} = 2 \in \mathbf{Z}.$$

Suppose that the statement is true for $k = n$. Our goal is to find a number N_{n+1} such that

$$\left(\sqrt{2}-1\right)^{n+1} = \sqrt{N_{n+1}} - \sqrt{N_{n+1} - 1}$$

and

$$\sqrt{2}\sqrt{N_{n+1}}\sqrt{N_{n+1} - 1} \in \mathbf{Z}.$$

From the induction hypothesis we have:

$$
\begin{aligned}
\left(\sqrt{2}-1\right)^{n+1} &= \left(\sqrt{2}-1\right)^n \left(\sqrt{2}-1\right) \\
&= \left(\sqrt{N_n} - \sqrt{N_n - 1}\right)\left(\sqrt{2}-1\right) \\
&= \left(\sqrt{N_n}\sqrt{2} + \sqrt{N_n - 1}\right) - \left(\sqrt{2}\sqrt{N_n - 1} + \sqrt{N_n}\right)
\end{aligned}
$$

1

Now observe that:

$$\left(\sqrt{N_n}\sqrt{2}+\sqrt{N_n-1}\right)^2 = 2N_n+(N_n-1)+2\sqrt{2}\sqrt{N_n}\sqrt{N_n-1}$$
$$= 3N_n-1+2\sqrt{2}\sqrt{N_n}\sqrt{N_n-1}$$
$$\in \mathbf{Z}.$$

Let K be the number $3N_n-1+2\sqrt{2}\sqrt{N_n}\sqrt{N_n-1}$.

$$\left(\sqrt{2}\sqrt{N_n-1}+\sqrt{N_n}\right)^2 = 2\left(N_n-1\right)+N_n+2\sqrt{2}\sqrt{N_n}\sqrt{N_n-1}$$
$$= 3N_n-2+2\sqrt{2}\sqrt{N_n}\sqrt{N_n-1}$$
$$= K-1\in\mathbf{Z}$$

Therefore we have:

$$\sqrt{K}-\sqrt{K-1} = \left(\sqrt{2}\sqrt{N_n}+\sqrt{N_n-1}\right)-\left(\sqrt{2}\sqrt{N_n-1}+\sqrt{N_n}\right)$$
$$= \left(\sqrt{2}-1\right)^{n+1},$$

and

$$\sqrt{K}\sqrt{K-1}\sqrt{2}$$
$$= \left(\sqrt{N_n}\sqrt{2}+\sqrt{N_n-1}\right)\left(\sqrt{2}\sqrt{N_n-1}+\sqrt{N_n}\right)\sqrt{2}$$
$$= 3\sqrt{N_n}\sqrt{N_n-1}\sqrt{2}+2N_n+2\left(N_n-1\right)$$
$$\in \mathbf{Z},$$

since $\sqrt{2}\sqrt{N_n}\sqrt{N_n-1}\in\mathbf{Z}$ by the induction hypothesis. So K has all the properties we want for N_{n+1}. Therefore, if we take $N_{n+1}=K$, we are done.

1.2. From PROBLEM 1.1.2 in the text, we know that:

$$1+2+3+\ldots+k = \frac{k^2+k}{2}.$$

We want to find:

$$1+3+5+7+\ldots+(2k-1).$$

(Note that the first integer is $1 = 2 \cdot 1 - 1$, the second is $3 = 2 \cdot 2 - 1$, so the kth will be $(2k - 1)$.)

Since adding all numbers from 1 to $(2k - 1)$ and then subtracting the even numbers in this sum gives the sum of the odd numbers from 1 to $(2k - 1)$, the last expression can be written as:

$$[1 + 2 + 3 + 4 + \ldots + (2k - 1)] - [2 + 4 + 6 + \ldots + (2k - 2)].$$

We have a formula for the first term. In the second term we can factor out 2. We obtain

$$[1 + 2 + 3 + 4 + \ldots + (2k - 1)] - 2 \cdot [1 + 2 + 3 + \ldots + (k - 1)].$$

Using the formula that was derived in the text we have that the last expression equals

$$\left[\frac{(2k - 1)^2 + (2k - 1)}{2} \right] - 2 \cdot \left[\frac{(k - 1)^2 + (k - 1)}{2} \right].$$

Doing some algebra we finally find

$$
\begin{aligned}
1 + 3 + \ldots + (2k - 1) & \\
&= \frac{(2k - 1)^2 + (2k - 1)}{2} - 2 \cdot \frac{(k - 1)^2 + (k - 1)}{2} \\
&= \frac{(2k - 1)((2k - 1) + 1)}{2} - 2 \cdot \frac{(k - 1)((k - 1) + 1)}{2} \\
&= \frac{(2k - 1) \cdot 2k - (k - 1) \cdot 2k}{2} \\
&= \frac{2k \left[(2k - 1) - (k - 1) \right]}{2} \\
&= \frac{2k^2}{2} \\
&= k^2.
\end{aligned}
$$

1.3. Let us first find a formula for the sum of the first k squares of integers. We will follow the same scheme as in PROBLEM 1.1.2 in the text. First observe that

$$\ell^3 - (\ell - 1)^3 = (\ell^3 - \ell^3 + 3\ell^2 - 3\ell + 1) = 3\ell^2 - 3\ell + 1.$$

Adding now from $\ell = 1$ to k we obtain a 'telescopic sum' on the left hand side in which most of the terms will cancel:

$$(k^3 - (k-1)^3) + ((k-1)^3 - (k-2)^3) + \cdots + (2^3 - 1) + (1 - 0)$$
$$= (3k^2 - 3k + 1) + (3(k-1)^2 - 3(k-1) + 1) + \cdots + (3 - 3 + 1).$$

Simplifying, reordering, and applying the formula for the sum of the first k integers, we obtain:

$$
\begin{aligned}
k^3 &= 3(k^2 + (k-1)^2 + \cdots + 1) \\
&\quad -3(k + (k-1) + \cdots + 1) + (1 + 1 + \cdots + 1) \\
&= 3(k^2 + (k-1)^2 + \cdots + 1) - 3\frac{k(k+1)}{2} + k \\
&= 3(k^2 + (k-1)^2 + \cdots + 1) - \frac{3k^2 + k}{2}.
\end{aligned}
$$

Finally, solving for $(k^2 + (k-1)^2 + \cdots + 1)$, we find:

$$(k^2 + (k-1)^2 + \cdots + 1) = \frac{2k^3 + 3k^2 + k}{6} \; .$$

For the sum of the cubes, we repeat this scheme again: we first observe that

$$\ell^4 - (\ell - 1)^4 = 1 \cdot (\ell^4 - \ell^4 + 4\ell^3 - 6\ell^2 + 4\ell - 1 = 4\ell^3 - 6\ell^2 + 4\ell - 1.$$

Adding now from $\ell = 1$ to k we obtain a 'telescopic sum' on the left hand side in which most of the terms will cancel:

$$[(k^4 - (k-1)^4] + [((k-1)^4 - (k-2)^4] + \cdots + [((2^4 - 1) + (1 - 0)]$$
$$= [(4k^3 - 6k^2 + 4k - 1] + [(4(k-1)^3 - 6(k-1)^2] + [(4(k-1) - 1]$$
$$+ \cdots + (4 \cdot 1^3 - 6 \cdot 1^2 + 4 \cdot 1 - 1).$$

Simplifying and reordering the last expression, we find:

$$k^4 = 4(1 + 2^3 + \cdots + k^3) - 6(1 + 2^2 + \cdots + k^2) + 4(1 + 2 + \cdots + k) - (k).$$

Solving for $(1 + 2^3 + \cdots + k^3)$, we find:

$$1^3 + 2^3 \cdots + k^3 = \frac{k^4 + 6(1 + 2^2 + \cdots + k^2) - 4(1 + 2 + \cdots + k) + (k)}{2}$$

$$= \frac{1}{2}\left\{ k^4 + 6\frac{2k^3 + 3k^2 + k}{6} - 4\frac{k^2 + k}{2} + k \right\}$$

$$= \frac{k^2(k+1)^2}{4} \quad , \text{ after simplifying.}$$

1.4. Let $\{a_1, a_2, \ldots, a_{52}\}$ be the collection. Using the euclidean algorithm, we can write each number a_k in the collection as $100 \cdot q_k + r_k$, with $1 \le r_k \le 100$.

For two different values of k, say k_1 and k_2, note that

$$a_{k_1} + a_{k_2} = 100 \cdot (q_{k_1} + q_{k_2}) + (r_{k_1} + r_{k_2})$$

and

$$a_{k_1} - a_{k_2} = 100 \cdot (q_{k_1} - q_{k_2}) + (r_{k_1} - r_{k_2}).$$

Thus, $a_{k_1} \pm a_{k_2}$ is a multiple of 100 if and only if $r_{k_1} \pm r_{k_2}$ is a multiple of 100. Therefore, it suffices to prove the statement for the collection $\{r_1, r_2, \ldots, r_{52}\}$.

Now, $\{r_1, r_2, \ldots, r_{52}\}$ forms a collection of numbers between 1 and 100. If two of them are equal, then their difference is divisible by 100, and we are done. Therefore we can assume that they are all distinct.

Let $s_k = 100 - r_k$. Then s_k is a collection of 52 distinct numbers between 0 and 99. Thus, $\{r_1, r_2, \ldots, r_{52}, s_1, s_2, \ldots, s_{52}\}$ forms a collection of 104 numbers between 0 and 100. By the pigeonhole principle, some number N must be repeated. A number cannot be repeated more than once, since the r_k's are distinct, and hence so are the the the s_k's. Since we have 104 'pigeons' and only 101 'holes', we can assume that $N \ne 50$.

Since N is repeated, there must be k_1, k_2 such that $N = s_{k_1} = r_{k_2}$. From the definition of the s_k's, we find

$$100 - r_{k_1} = r_{k_2},$$

which implies

$$100 = r_{k_1} + r_{k_2}.$$

Since $r_{k_2} = N \neq 50$, we must have $r_{k_1} \neq r_{k_2}$. Thus, we have shown that two of the r_k's add up to a multiple of 100, as desired.

1.5. Write the equation as

$$n(m - 1) = m.$$

This means that $m - 1$ divides m. But this can only happen if $m = 2$ or if $m = 0$.

If $m = 2$, the equation above reads

$$n(2 - 1) = 2,$$

which implies $n = 2$.

If $m = 0$, the equation above reads

$$n(0 - 1) = 0,$$

which implies $n = 0$.

Thus, the only solutions are $m = n = 0$ and $m = n = 2$.

1.6. We have to find the number of times that 10 divides the number 200!. Since $10 = 2 \cdot 5$, we can just count how many times 2 and 5 divide 200!, and then take the least of these.

We can count things in the following way:

Between 1 and 200, there are:

100	multiples of 2
50	multiples of 4
25	multiples of 8
12	multiples of 16
6	multiples of 32
3	multiples of 64
1	multiple of 128
TOTAL	197

Between 1 and 200, there are:

40	multiples of 5
8	multiples of 25
1	multiple of 125
TOTAL 49	

Since 49 is less than 197, we have that 49 zeros end the number 200!.

The reason this method works can be explained as follows: suppose that we write all the numbers between 1 and 200 in a line. Put three dots arranged vertically over each multiple of 125, two arranged vertically over each multiple of 25 and one over each multiple of 5. In this way we have dots at three different levels. The total number of dots is the number of times that 5 divides 200!. If now, as we did above, we count the multiples of 5, we are counting all the dots that are at the first level, when we count the multiples of 25 we are counting the number of dots at the second level, etc. At the end we find the total number of dots, which is, as we said, the number of times that 5 divides 200!.

1.7. Proceed as follows:

$$
\begin{aligned}
2^{300} \cdot 5^{600} \cdot 4^{400} &= 2^{300} \cdot 5^{600} \cdot 2^{800} \\
&= 2^{600} \cdot 5^{600} \cdot 2^{500} \\
&= 10^{600} \cdot 2^{500}.
\end{aligned}
$$

Therefore, the number $2^{300} \cdot 5^{600} \cdot 4^{400}$ ends in 600 zeros.

1.8. The only thing that can be said about the number of people who shake hands an even number of times is that it has the same parity as the total number of people that shook hands. This is just because, as it is proved in the text, the number of people who shake hands an odd number of times is even. Besides that, any combination can occur: suppose that M people in a group will shake hands. Let N be any number less than M and with the same parity as M. Then one can find a combination in which exactly N people will shake hands an even number of times.

1.9. Between 1 and 100, we have:

$$
\begin{array}{rlcr}
9 & \text{numbers with 1 digit} & = & 9 \text{ digits} \\
90 & \text{numbers with 2 digits} & = & 180 \text{ digits} \\
1 & \text{number with 3 digits} & = & 3 \text{ digits} \\
& \text{TOTAL} & & \overline{192 \text{ digits}}
\end{array}
$$

1.10. If $k!$ does not end in a zero, that means that 10 does not divide it, or equivalently that not both 2 and 5 divide it. This leaves only five possibilities: 0,1,2,3 and 4.

1.11. Suppose that we have a number N with k digits, which we write as $N = a_k a_{k-1} \ldots a_1 a_0$. We can also write N as:

$$
\begin{aligned}
N &= a_k 10^k + a_{k-1} 10^{k-1} + \cdots + a_1 10 + a_0 \\
&= a_k (\underbrace{99\cdots 9}_{k\,\text{digits}} + 1) + a_{k-1}(\underbrace{99\cdots 9}_{k-1\,\text{digits}} + 1) + \cdots + a_1(9+1) + a_0 \\
&= [a_k \underbrace{99\cdots 9}_{k\,\text{digits}} + a_{k-1}\underbrace{99\cdots 9}_{k-1\,\text{digits}} + \cdots + 9] + [a_k + a_{k-1} + \cdots + a_0]
\end{aligned}
$$

Since both N and the first term are divisible by 9, the second term, namely $a_k + a_{k-1} + \cdots + a_0$, must also be divisible by 9. Also, the number $a_k + a_{k-1} + \cdots + a_0$ has fewer digits than the original number N that we started with. When we add the digits of a number divisible by 9 we obtain a number with fewer digits than the original that is also divisible by 9. If we continue this process, at some point we will obtain a number with only one digit that is divisible by nine. This number has to be 9.

1.12. The point is that after the process we always end up with the letter d. Here is why: Let the chosen integer be k. $9 \cdot k$ has 2 digits and is divisible by 9. When we add its digits together we obtain a number that has fewer digits than $9 \cdot k$ (so it has only one digit) and is also divisible by 9. So the number we obtain must be 9 itself. If we now subtract 5 we obtain 4, which corresponds to the letter d. A problem may arise, though, if your friend was born in the Dominican Republic and likes Owls. . .

1.13. Let us proceed using the same strategy as in Exercise 6. Suppose that we write out all the numbers that we are multiplying to form

$n!$, i.e. $1 \cdot 2 \cdot 3 \cdots n$. Put k dots, arranged vertically, over each number of the string that is divisible by p^k until we exhaust all numbers in the string. It is clear that the number of factors of p in $n!$ equals the number of dots, so let us count the dots.

When we count how many numbers are divisible by p it is as if we were counting the bottom dots. If we now count how many numbers are divisible by p^2, we are counting the dots 'at the second level'. Proceeding this way, we will eventually reach some k such that no numbers in the string are divisible by p^k (when $p^k > n$), i.e. we have exhausted all the 'levels'.

Therefore, the number of factors of p that occur in $n!$ equals the number of numbers between 1 and n that are divisible by p plus the number of numbers between 1 and n that are divisible by p^2, etc. Now, how many numbers are divisible by p^k, some k, between 1 and n? We will have that $1 \cdot p^k$, $2 \cdot p^k$, ..., $l \cdot p^k$ are the ones that are divisible by p^k, up to the greatest l such that $l \cdot p^k$ is less than or equal to n. This implies that between 1 and n there are exactly

$$\left[\frac{n}{p^k} \right]$$

numbers divisible by p^k, where $[x]$ denotes the greatest integer lesser than x.

This is because

$$p^k \cdot \left[\frac{n}{p^k} \right] < n,$$

and

$$p^k \cdot \left(\left[\frac{n}{p^k} \right] + 1 \right) > n,$$

so that $\left[\dfrac{n}{p^k} \right]$ is the greatest l such that $l \cdot p^k$ is less than or equal to n. Thus, the final formula is

$$\sum_{k=1}^{\infty} \left[\frac{n}{p^k} \right].$$

(Note that we can safely sum to ∞ since $\left[\dfrac{n}{p^k}\right]$ equals 0 for all k sufficiently large.)

1.14. 99% of 500 pounds is 495 pounds of water. So we have 5 pounds that are not water. Once the watermelon has dried out for a while, these 5 pounds represent a 2% of the total weight of the watermelon (since water is 98% of the weight of the watermelon). If 5 pounds is 2%, then 250 pounds is 100%. Thus, the watermelon now weighs 250 pounds.

1.15. The total number of games played, if each team plays every other exactly once, is $14 + 13 + 12 + \cdots + 2 + 1 = 105$. To see this, count first the games that the best team played (14), then the games that the second best team played, excluding the game with the best team—that we have already counted—(13), etc. Each game generates a total of 4 points, shared between the two teams in that game. Thus the total number of points is $4 \cdot 105 = 420$.

Now, if every team ends up with a different total score and the team with lowest total scored 21, it must be that the team with second lowest total scored at least 22, the one with third lowest total scored at least 23, up to the best team which scored at least 35 points. Adding up the total scores we must have that the total number of points has to be greater than or equal to

$$21 + 22 + 23 + \cdots + 34 + 35 = 420.$$

Since this is exactly the total number of points, it must be that the team with second lowest total scored exactly 22, the one with third lowest total scored exactly 23, and the best team scored exactly 35 points (note that if any team had scored more, then the sum of the final scores of all the teams would exceed 420, which is impossible).

The maximum number of points that a team could have received is

$$3 \text{ points} \times 14 \text{ games} = 42 \text{ points}.$$

Thus the best team lost only 7 points. Note that, for each loss, 2 points are subtracted (a win is 3 but a loss is 1). If the best team had not had any draws, then the total number of points lost would be even, never 7. Therefore one of the games of the best team was surely a draw.

1.16. The correct number is 0.

1.17. The correct number is $2 - 2k$.

1.18. Suppose that the assertion is false. Consider an equilateral triangle of side 1.

If the lower vertex is red, then the other two must be blue and yellow:

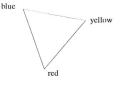

Now erect a second triangle as shown below. The new vertex must be red:

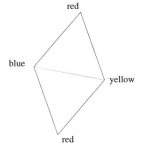

Rotate the figure through an angle θ chosen so that the upper right vertex moves one unit:

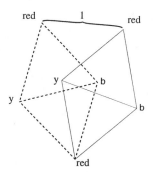

Reasoning as before forces the upper vertex of the new dotted figure to be red. But that vertex is at distance one from the upper vertex of the old figure. Thus we have found a segment of length 1 with both ends red.

1.19. One can color the plane in the following way. The hexagons are regular and have diameter 1. Each letter stands for a color. We color the inside and the left half of the boundary of each hexagon, including the top vertex and excluding the bottom one, with the color corresponding to the letter. Notice that the distance between two hexagons of the same color is always greater than one.

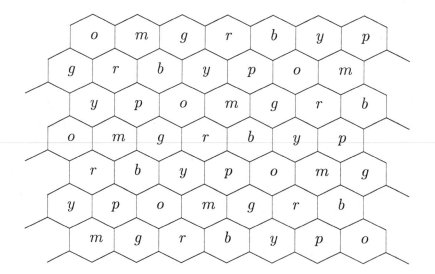

1.20. If there is only one unfaithful husband, then all the wives will know about it except for one—the cheated wife. She will know right away that she is the cheated one, since she knows of no unfaithful husbands and yet there is one, so it has to be hers. That day her husband will be seen with a big "A" in his forehead.

Now, if there are two unfaithful husbands, there must be two women who know only about one unfaithful husband. But they know that, if there is only one unfaithful husband, then on the same day some guy has an "A" in his forehead. If it turns out that nobody had such ill luck, then it must be that there are at least two unfaithful husbands. Thus these two women will know that their husbands were the unfaithful ones. The next day—the wives would wait overnight just in case they missed the husband with the tattooed forehead—they will know that for sure they are the cheated wives, and they will tattoo their husbands' foreheads.

Suppose now that a wife knows of k unfaithful husbands. If after $k - 1$ days there are no tattooed foreheads, that means that the actual number of cheated husbands must be $k + 1$, so that wife is doomed to tattoo a big "A" on her husband's forehead.

This means that after a finite number of days—less than the number of couples in the town—everything will be known.

1.21. If the mayor announces how many husbands were unfaithful, everything will be clear right away: all the wives that know about fewer unfaithful husbands than the number given by the mayor will immediately know that they have been cheated.

1.22. If we glue the sides with no twists we obtain a torus. The number γ satisfying $V - E + F = \gamma$ for the Klein bottle is 0, the same as for the torus.

1.23. Note that

$$\left[\frac{1}{2!} + \frac{2}{3!} + \cdots + \frac{n}{(n+1)!}\right] + \left[\frac{1}{2!} + \frac{1}{3!} + \cdots + \frac{1}{(n+1)!}\right]$$

$$= \left[\frac{2}{2!} + \frac{3}{3!} + \cdots + \frac{n+1}{(n+1)!}\right]$$

$$= \left[\frac{1}{1!} + \frac{1}{2!} + \cdots + \frac{1}{n!}\right].$$

Therefore we have:

$$\left[\frac{1}{2!} + \frac{2}{3!} + \cdots + \frac{n}{(n+1)!}\right]$$

$$= \left[\frac{1}{1!} + \frac{1}{2!} + \cdots + \frac{1}{n!}\right] - \left[\frac{1}{2!} + \frac{1}{3!} + \cdots + \frac{1}{(n+1)!}\right]$$

$$= 1 - \frac{1}{(n+1)!}$$

1.24. To count the total number of subsets of a set S with k elements, we can count the number of subsets of S that have a given number of elements n, and then add up from $n = 0$ to k:

$$
\begin{aligned}
N(n) &= \{\text{number of subsets with } n \text{ elements}\} \\
&= \{\text{number of ways to choose } n \text{ elements} \\
&\qquad \text{out of a set with } k \text{ elements}\} \\
&= \binom{k}{n}.
\end{aligned}
$$

Thus we only have to show

$$\binom{k}{0} + \binom{k}{1} + \cdots + \binom{k}{k} = 2^k.$$

To prove this, apply the binomial theorem to $(1+1)^k$.

Thus, S has $N(0) + N(1) + \cdots + N(k) = 2^k$ subsets.

1.25. For the first strategy, the probability of winning is:

$$\frac{\text{number of winning cases}}{\text{number of possible cases}} = \frac{a}{a+b}.$$

To find the probability of winning with the second strategy, we have to take into account the fact that the probability of drawing a white ball in the second draw depends on which ball was thrown away in the first draw. We have:

Pr{ drawing a white ball at the end }

$$= \mathbf{Pr}\{\text{white at the end}\,|\,\text{assuming 1}^{\text{st}}\text{ white}\} \cdot \mathbf{Pr}\{1^{\text{st}}\text{ white}\}$$
$$+\mathbf{Pr}\{\text{white at the end}\,|\,\text{assuming 1}^{\text{st}}\text{ black}\} \cdot \mathbf{Pr}\{1^{\text{st}}\text{ black}\}$$
$$= \frac{a-1}{a+b-2} \cdot \frac{a}{a+b} + \frac{a}{a+b-2} \cdot \frac{b}{a+b}$$
$$= \frac{a^2 - a + ab}{(a+b)(a+b-2)}$$
$$= \frac{a(a+b-2) + a}{(a+b)(a+b-2)}$$
$$= \frac{a}{a+b} + \frac{a}{(a+b)(a+b-2)}.$$

It is similar to the Monty Hall problem: player B eliminates one of the black balls, just like Monty Hall would eliminate one of the doors in the game.

1.26. The answer is "no":

Color the cube in a checker-board fashion, as in the figure:

Note that we have $5 + 4 + 5 = 14$ black cubes and $4 + 5 + 4 = 13$ white cubes. Note also that the center cube is white.

Starting at any cube, we want the termite to finish in the center cube. Note that no subcubes of the same color have a common face. Therefore the termite will always go from a white cube to a black cube and from a black cube to a white cube.

Therefore, the sequence of the colors of the cubes in the termite journey will be of the form $\dots b\,w\,b\,w\,b\,w$. This is, excluding the initial sub-cube we have

$$\underbrace{b\,w\,b\,w\,\ldots b\,w}_{\text{26 times}}.$$

Thus the first cube should also have been white. This would make a total of 14 white cubes, and we only have 13.

This shows that it is impossible for the termite to eat the whole cube and finish at the middle one.

1.27. In all the cases asked it is possible to tile the floor—and actually not too hard. If two adjacent corners are missing, then tile the entire floor placing the tiles parallel to the side that has the missing corners. If the omitted squares are adjacent to each other, then place the tiles parallel to these two adjacent squares; then filling up the rest will be straightforward.

Chapter 2

A Deeper Look at Geometry

2.1.   a) For example, the complete graph on four vertices in the sphere requires 4 colors.

b) The complete graph on seven vertices in the torus requires 7 colors, as is shown in the figure below. We picture the torus as a rectangle in which we identify the top edge with the bottom edge and the left edge with the right edge. The letters a, a', b, b' in the figure indicate this identification.

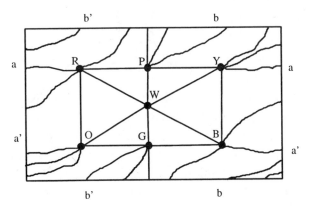

c) The chromatic number of a torus with two holes is 8.

2.2. Two colors are actually enough to color a map of this sort. The reason is that, in such a map, if one region has a common border with two other regions, then these two regions do not share a

17

common border. Here is a constructive procedure to color a map
of this kind. Let us use the colors red and purple, denoted by R
and P in the figure below. Start by coloring purple the portion
of the plane not intersected by any of the circles. Then color red
the portion of the interior of each circle that does not intersect
any other circle. Then color purple the regions that are the intersection of exactly two circles. Then color red the regions that
are the intersection of exactly three circles. Continue this way.
At the end we have a map in which no two adjacent regions are
the same color.

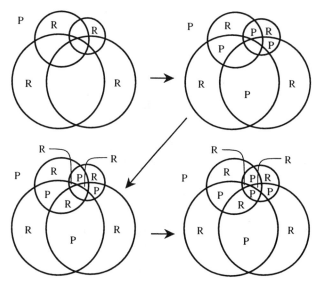

This procedure will work for any configuration of circles. Note
that, in this procedure, we are coloring purple the regions that are
the intersection of an even number of circles, and we are coloring
red the regions that are the intersection of an odd number of
circles. Suppose that we are in a region that is in the intersection
of k circles. If we cross a boundary of one such region, we will
either leave one of the k circles (so that now we are in a region
which is in the intersection of $k - 1$ circles) or enter a new circle
(so that now we are in a region which is in the intersection of
$k + 1$ circles). This is because we cannot leave two circles at the
same time, since if two circles have an arc in common, they must
coincide. Thus, a region that is in the intersection of k circles will

only have common edges with regions that are in the intersection of either $k-1$ or $k+1$ circles. But since k and $k-1$ have different parity, they are not the same color, and the same thing happens with k and $k+1$. Thus, the above procedure to color a map of this sort will always work.

2.3. View the triangle lying on one of its sides, as in the figure:

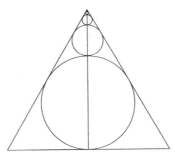

If we draw a vertical segment from the base of the triangle to the top vertex, then the length of this segment will be exactly the sum of the diameters of all the circles in the figure. Since the altitude of the triangle is 3, we have that the sum of the lengths of the diameters of all the triangles is 3, which implies that the sum of their radii is 1.5. Now, since we are doing this process in every vertex of the triangle, the sum of the radii of all the circles involved is $3 \cdot 1.5 - 2 = 2.5$. Notice that we are subtracting 2 because, otherwise, the radius of the circle in the center (which has length 1) would be counted thrice. Thus, the answer is 2.5.

2.4. Divide the square into 4 smaller sub-squares. These squares have diagonal $\sqrt{2}/2$. Therefore, if we want to arrange the points so that all points are at a distance greater than $\sqrt{2}/2$ from each other, it follows that no two points can lie in the same sub-square. But we have only 4 sub-squares in which to place 5 points. By the pigeonhole principle, two points must lie in the same sub-square, and therefore, the distance between these two points must be less than or equal to $\sqrt{2}/2$.

2.5. View the triangle as lying on its longest side, denoted by a, and let b and c be the left and right sides respectively, as in the figure below:

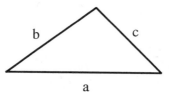

We are assuming that $a = n$. If $b = k$, for k an integer between 1 and n (both inclusive), then c can be $(n-k+1), (n-k+2), \ldots, n$ (note that we must have $a < b + c$, otherwise the figure cannot be a triangle). Thus, if $b = k$, we have k choices for c. This gives a total of

$$1 + 2 + 3 + \cdots + n = \frac{n(n+1)}{2} \quad \text{triangles.}$$

Note, though, that the isosceles triangles for which two of the sides have length n are congruent, so they have been counted twice (namely, for $a = b = n$, $c = k < n$ and for $a = c = n$, $b = k < n$). Since there are $n - 1$ such triangles, the final answer is

$$\frac{n(n+1)}{2} - (n-1) = \frac{n^2 - n + 2}{2} \quad \text{non-congruent triangles.}$$

2.6. Denote the sides of the triangle by a, b, c where a is the hypotenuse. The equations we have are:

$$a^2 = b^2 + c^2$$
$$a + b + c = 60$$
$$12a = bc \qquad \text{from similarity of triangles.}$$

Solving these 3 equations (which is a standard but tedious computation) we obtain that the hypothenuse is $a = 25$ inches and the other sides are $b = 20$ inches and $c = 15$ inches respectively.

2.7. To find the proportion of the plane covered in the square packing, just divide the area of the inscribed circle (of diameter 1) by the area of the square of side 1 in which the circle is inscribed. This gives $\pi/4$.

In the case of the hexagon, divide the area of the inscribed circle (of radius $\sqrt{3}/2$) by the area of a hexagon of side 1 in which the circle is inscribed. This gives $\pi/(2\sqrt{3})$.

Note that the second packing is more efficient.

2.8. To see how to tile the plane by hexagons, see the figure in the solution of Exercise 19 of Chapter 1. One cannot tile the plane by pentagons because the angle between two adjacent sides is $3\pi/5$. If we put several pentagons with a common vertex and a common side—i.e.
if we try to pack them—we will need more than 3 pentagons, since $9\pi/5 < 2\pi$, but less than 4 pentagons, since $12\pi/5 > 2\pi$. This is not possible.

2.9. Adjoining two congruent triangles along a corresponding side, we obtain parallelograms. The plane can always be tiled by parallelograms of any shape (see Figure 3 in the text), so the answer is affirmative.

2.10. First, by rescaling by the least common multiple of the denominators of the rational lengths of the sides, we can assume, without loss of generality, that the sides of the rectangle have integer lengths. Call these lengths a and b. Let p and q be such that $p \cdot a = q \cdot b$. Then, to tile an infinite horizontal strip of height $p \cdot a$, we can combine blocks of q rectangles piled up, each with vertical side of length b and blocks of p rectangles piled up, each with vertical side of length a: in either way we get the same height, $p \cdot a$ (see the figure below). Since the number of such combinations is infinite, any infinite horizontal strip of height $p \cdot a$ can be tiled in infinitely many ways, and then we can pile up strips to cover the entire plane.

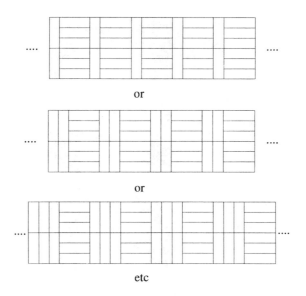

2.11. False: an equilateral triangle of side d has diameter d, but it cannot be inscribed in a circle of diameter d, since the distance from the center of the triangle to each vertex is $d/\sqrt{3}$, which is greater than the radius of the circle.

2.12. A set of width d need not have diameter d: a strip of width d, for example, has infinite diameter. On the other hand, a set of diameter d always has width d: if we arrange a strip so that some segment of length d joining two points of the set (which must exist since the set has diameter d) is perpendicular to the strip, then any point outside of the strip will be at distance greater than d from one of the points.

2.13. Convexity can be defined as follows: a set is convex if, for any two points p, q in the set, all the points of the form

$$p \cdot t + q \cdot (1 - t), \quad 0 \le t \le 1$$

also lie in the set.

Let $p, q \in X + Y$, and let $0 \le t \le 1$. We want to show that $p \cdot t + q \cdot (1 - t) \in X + Y$. By definition of $X + Y$, we can write $p = p_X + p_Y$ and $q = q_X + q_Y$, with $p_X, q_X \in X$ and $p_Y, q_Y \in Y$. Since X is convex, the point $p_X \cdot t + q_X \cdot (1 - t)$ lies in X, and since Y is convex, the point $p_Y \cdot t + q_Y \cdot (1-t)$ lies in Y. Thus, we have

$$
\begin{aligned}
p \cdot t &+ q \cdot (1-t) \\
&= (p_X + p_Y) \cdot t + (q_X + q_Y) \cdot (1-t) \\
&= (p_X \cdot t + q_X \cdot (1-t)) + (p_Y \cdot t + q_Y \cdot (1-t)) \in X + Y
\end{aligned}
$$

Therefore, $X + Y$ is convex.

The only thing that can be said about the diameter is that it is at least $d\sqrt{2}$ and at most $2d$. To show this, use the triangle inequality. These bounds are sharp: taking for example two discs of diameter 1, the sum will be a disc of diameter 2. On the other hand, taking two perpendicular segments of length d, their sum will be a square of side d, which has diameter $d\sqrt{2}$.

Concerning the width, nothing can be said. For example, taking X as a horizontal infinite stripe of diameter d and Y as a vertical infinite stripe of diameter d, $X + Y$ will be the whole plane, which has infinite diameter.

2.14. The area of the sum of two sets X, Y will be not less than the area of each of the sets since, if $q \in Y$, $X + q$ has already the same area as X, and the same reasoning applies for Y. In fact, we can be more precise: there is an inequality, called the *Brunn-Minkowski* inequality, that states

$$
\sqrt{(\text{Area}(A))} + \sqrt{(\text{Area}(B))} \leq \sqrt{(\text{Area}(A + B))}.
$$

It is hard to prove this inequality rigorously, but the idea is the following: first check that, when A and B are rectangles such that each side of A is parallel to a side of B, then the inequality holds. This is easy to prove: first, if A is a rectangle of dimensions (a, b) and B is a rectangle of dimensions (c, d), then $A + B$ is a rectangle of dimensions $(a + c, b + d)$, as illustrated in the following picture:

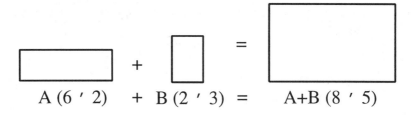

A (6 ′ 2)   +   B (2 ′ 3)   =   A+B (8 ′ 5)

Now, the area of A is ab, the area of B is bc, and the area of $A+B$ is $(a+c)(b+d)$. We need to prove

$$\sqrt{ab} + \sqrt{cd} \le \sqrt{(a+c)(b+d)}.$$

Squaring both sides we obtain

$$ab + cd + 2\sqrt{ab}\sqrt{cd} \le (a+c)(b+d).$$

Simplifying, we find

$$2\sqrt{ab}\sqrt{cd} \le ad + cb.$$

Squaring both sides again, we obtain

$$4abcd \le (ad)^2 + (cb)^2 + 2abcd.$$

Substracting $4abcd$ from both sides, we find

$$0 \le (ad)^2 + (cb)^2 - 2abcd = (ad - bc)^2,$$

which is definitely true. Since the process can be reversed, we are done.

The area of any set in the plane can be approximated by the sum of the areas of very small rectangles that form a partition of the set, as in the following figure:

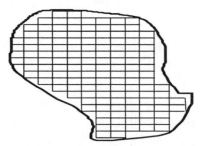

Finally, one expresses the area of the sets A and B as a limit of sums of areas of small rectangles and uses the fact that the inequality holds for rectangles. The details are quite technical, and will not be presented here.

We encourage the reader to experiment with different kinds of sets, checking that the Brunn-Minkowski inequality holds in each case. Here is an interesting example: the area of a line is 0, the area of a disc of radius 1 is π, but the area of a line plus a disc of radius 1 is infinity.

2.15. The area gets multiplied by $2^2 = 4$. To show this think first of the fact that, when we multiply by 2, the sides of any square in the plane doubles, and so its area quadruples. Since squares are the building blocks to measure area (i.e. to find the area of a set we tile it as best as we can with very small squares and then we add the areas of the squares), the area of any set gets multiplied by 4.

Note that the original position of the set does not play any role. When we multiply by 2, if the set is originally far from 0 it will be now twice as far, but its shape will be the original one rescaled by 2.

2.16. In this case the area is not invariant under the geometric operation of translation. This is, if the original set S was very far outside the circle of center 0 and radius 1, then the area of S' will be now very small. For example, if S is the part of the plane that lies outside a circle of radius $R_1 = 100$ centered at 0 and inside a circle of radius $R_2 = 1,000$ centered at 0, then S' will be the region inside a circle of radius 0.01 and outside a circle of radius 0.001, both centered at 0. The area of S is $(1,000,000 - 1,000)\pi = 999,000\pi$, whereas the area of S' is $(0.0001 - 0.000001)\pi = 0.000099\pi$.

2.17. For subsets of the line, a set is convex if and only if it contains all the points between two given points, i.e. if it is connected. The sum of two connected sets (here, connected means that it has no 'holes') is clearly connected.

The diameter and the width are the same quantity for subsets of the line. For subsets of the line, the diameter of the sum of two sets is the sum of the diameters of the sets.

2.18. The idea is as follows. Start with the needle in vertical position. Moving the needle almost vertically, describing a long almost flat arc we end up with the needle slightly rotated an angle θ and in a different location. Repeat the same process moving the needle in the opposite direction along an arc similar to the previous one but with opposite curvature. At the end of this step we have the needle rotated an angle 2θ. After doing enough steps (and choosing the angle θ properly) we rotate the needle through $180°$. The ink blot will have the shape of a star with very small kernel and very thin and long beams. That is the idea of why the final area can be as small as we want: very thin long cusps have very small area.

The actual solution is technically more complicated. For more details, see the article *The Kakeya Problem* by A. S. Besicovitch, in the journal American Mathematical Monthly, # 70 (1968), pages 697-712. The article in the same subject by F. Cunningham JR, in the same journal, # 78 (1971), pages 114-130 is also recommended to the interested reader.

2.19. Denote the bigger angle by α and the smaller by β. Then we have $\sin\alpha = 1/\sqrt{5}$, $\cos\alpha = 2/\sqrt{5}$, $\sin\beta = 1/\sqrt{10}$ and $\cos\beta = 3/\sqrt{10}$. This implies

$$
\begin{aligned}
\sin(\alpha + \beta) &= \sin\alpha\cos\beta + \sin\beta\cos\alpha \\
&= \frac{1}{\sqrt{5}}\frac{3}{\sqrt{10}} + \frac{2}{\sqrt{5}}\frac{1}{\sqrt{10}} \\
&= \frac{3}{5\sqrt{2}} + \frac{2}{5\sqrt{2}} \\
&= \frac{1}{\sqrt{2}} \\
&= \sin 45°.
\end{aligned}
$$

Therefore, $\alpha + \beta = 45°$.

2.20. Denote the vertices of the quadrilateral by A, B, C, D so that A is opposite to C and B is opposite to D. Denote by N the point where the diagonals intersect. Finally let $\alpha = \angle ANB$ and $\beta = \angle BNC$, so that $\alpha + \beta = 180°$, and $\cos \beta = -\cos \alpha$. All this notation is depicted in the figure below.

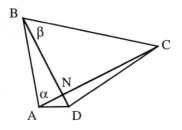

Apply the theorem of the cosine to the four triangles formed by the two intersecting diagonals:

$$
\begin{aligned}
|AB|^2 &= |AN|^2 + |BN|^2 - 2|AN||BN|\cos\alpha \\
|CD|^2 &= |CN|^2 + |DN|^2 - 2|CN||DN|\cos\alpha \\
|AD|^2 &= |AN|^2 + |DN|^2 - 2|AN||DN|\cos\beta \\
|CB|^2 &= |CN|^2 + |BN|^2 - 2|CN||BN|\cos\beta.
\end{aligned}
$$

Using $\cos\beta = -\cos\alpha$ we obtain

$$
\begin{aligned}
|AB|^2 &= |AN|^2 + |BN|^2 - 2|AN||BN|\cos\alpha \\
|CD|^2 &= |CN|^2 + |DN|^2 - 2|CN||DN|\cos\alpha \\
|AD|^2 &= |AN|^2 + |DN|^2 + 2|AN||DN|\cos\alpha \\
|CB|^2 &= |CN|^2 + |BN|^2 + 2|CN||BN|\cos\alpha.
\end{aligned}
$$

Add the first two equations and subtract the last two. Note that all the terms that are squared on the right hand side of the equations cancel:

$$
\begin{aligned}
&|AB|^2 + |CD|^2 - |AD|^2 - |CB|^2 \\
&= -2(|AN||BN| + |CN||DN| + |AN||DN| + |CN||BN|)\cos\alpha.
\end{aligned}
$$

Thus we have

$$|AB|^2 + |CD|^2 = |AD|^2 + |CB|^2 \ \text{ if and only if } \ \alpha = 90°,$$

which is what we wanted to prove.

2.21. We can assume that the vertices of the triangle lie in the boundary of the polygon. The perimeter of the polygon is the sum of the lengths of the sectors of the polygon joining each pair of vertices of the triangle. Each of these lengths is no less than the length of the corresponding side of the triangle. Therefore the sum of the lengths of the sectors of the polygon joining each pair of vertices of the triangle is greater than or equal to the sum of the lengths of the corresponding sides of the triangle, which is exactly the perimeter of the triangle.

2.22. Using the formula found in **PROBLEM 2.2.5** for the area of the triangle in terms of its sides we obtain:

$$6 = \sqrt{\frac{(3n+3)}{2} \cdot \frac{(n-1)}{2} \cdot \frac{(n+1)}{2} \cdot \frac{(n+3)}{2}}.$$

Solving this equation we find $n = 3$. Therefore the lengths of the sides are 3,4,5. One of the angles is a right angle. The cosines of the other two are 3/5 and 4/5.

2.23. Let the vertices of the triangle be denoted by A, B, C. Let us find the set of points N such that $\angle NAB = \angle NBC$. Given an angle θ, let N_θ be a point such that $\angle N_\theta AB = \angle N_\theta BC = \theta$. To find this point, draw a line through A at an angle θ from the segment AB and draw a line through B at an angle θ from the segment BC. These lines are not parallel (if they were, it is easy to see that the angle $\angle ABC$ would be 180°). The point N_θ is the point of intersection of both lines (note that since N_θ is in the first line, $N_\theta AB = \theta$ and since N_θ is in the second line, $N_\theta BC = \theta$). This is illustrated in the figure below.

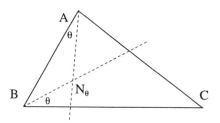

As we change θ, N_θ will describe a convex arc, as in the figure below:

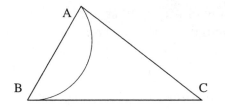

Similarly, the set of points M such that $\angle MBC = \angle MCA$ is a concave arc, as in the following figure:

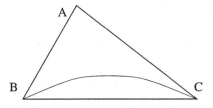

The first arc is tangent to BC, whereas the second is transversal. Since the first arc ends at A and the second ends at C, the arcs must intersect, as in the figure:

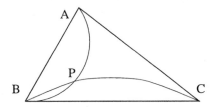

Let us denote the point of intersection by P. Since P is in the first arc, $\angle PAB = \angle PBC$, and since P is in the second arc, $\angle PBC = \angle PCA$. Thus, $\angle PAB = \angle PBC = \angle PCA$, as desired.

2.24. Each new row has two triangles more than the previous one. Since the first row has only one triangle, the number of triangles in a figure with n rows must be

$$1 + 3 + 5 + \cdots + 2n - 1.$$

This sum equals n^2 (cf. Exercise 2, Chapter 1).

Another way to look at it is comparing areas: the figure with n rows has base $n \cdot b$, and height $h \cdot n$, where b and h are the base and the altitude of the small triangles. Thus the area of the figure with n rows equals n^2 times the area of a small triangle.

2.25. It can be solved using analytic geometry: fix two coordinate axes and write everything explicitly. A nicer way is the following: denote the sides by p, q, r, and let a, b, c be the distances from P to the sides p, q, r respectively. Suppose that the triangle is laying on its side r. Let h be the altitude of the triangle.

Draw a line through P parallel to the side p. We obtain a smaller triangle $T1$, whose altitude is $h - a$. Then draw a line through P parallel to q. We obtain an even smaller triangle $T2$ inside $T1$ whose altitude is the altitude of $T1$ minus b, i.e. $h - a - b$. On the other hand, the top vertex of $T2$ is P, and the base lies on r. Therefore the altitude of $T2$ is equal to the distance from P to r, which is equal to c. Thus we have $h - a - b = c$, which is what we wanted to prove.

2.26. See the figure below:

1) The points corresponding to valid triangles lie in the triangle of vertices $(0, 1)$, $(1, 1)$, $(1/2, 1/2)$.

2) The isosceles triangles correspond to points in the set of valid triangles that satisfy $x = y$. Thus it is the segment from $(1/2, 1/2)$ to $(1, 1)$.

3) To be equilateral, it must be that $x = y = 1$. Thus, the equilateral triangle corresponds to the point $(1, 1)$.

4) Right triangles satisfy $x^2 + y^2 = 1$. Thus they correspond to the arc of the unit circumference laying inside the set of admissible triangles.

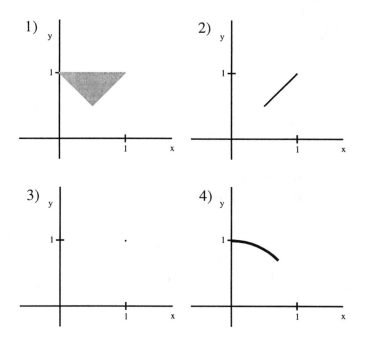

2.27. Assume that m is the hypotenuse. Then we must have $m^2 = \ell^2 + 100$, or $m^2 - \ell^2 = 100$. This can be written as $(m+\ell)(m-\ell) = 100$. Denote $m + \ell$ by a and $m - \ell$ by b. Then a and b divide 100 and their product is 100. Also, we have $m = (a+b)/2$ and $\ell = (a-b)/2$, so for m and ℓ to be integers we must have that a and b have the same parity, that is, they are either both odd or both even. Inspecting all the divisors of 100 we find that the only possibilities are $a = b = 10$ and $a = 50$, $b = 2$. This corresponds to $m = 10$, $\ell = 0$, which is the degenerate case, and $m = 26$, $\ell = 24$.

2.28. It is a general fact about triangles that the distance from a vertex to the point of intersection of the medians is 2/3 of the length of the corresponding median. To show this, let A, B, C denote the vertices, P, Q, R the mid-points of the respective opposite sides and X the point where the medians intersect, as in the following figure:

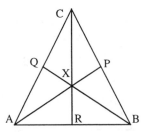

Draw the triangle $\triangle PQR$. It is similar to $\triangle ABC$ but half its size, so all distances are divided by 2. In particular, the length of the median corresponding to the vertex P of $\triangle PQR$ is $|AP|/2$. This implies that

$$\frac{|AP|}{|AX|} = \frac{|AP|/2}{|AP| - |AX|}.$$

Solving for $|AX|$ we find $|AX| = (2/3)|AP|$, which is what we wanted.

Once we have proved this fact, we can show that if the medians of two triangles are equal, then the triangle is isosceles. With the same notation as before, assume that the two equal medians are AP and BQ. Then we must have $|AX| = (2/3)|AP| = (2/3)|BQ| = |BX|$. This implies that the triangle $\triangle ABX$ is isosceles, which implies that the angles $\angle ABQ$ and $\angle BAP$ are equal. Since the triangles $\triangle ABQ$ and $\triangle BAP$ have the side AB in common, the sides AP and BQ have the same length and the angle in common also has the same measure. Therefore the two triangles must be congruent. Finally, this implies that $\angle ABC = \angle BAC$, which implies that the triangle $\triangle ABC$ is isosceles.

2.29. Assume that one of the axes of symmetry is horizontal. Denote the other axis by ℓ, and denote the angle between the axes by α. If we apply the symmetry along the horizontal axis first, the other axis will move to form an angle $-\alpha$ with the horizontal. This means that the line m that forms an angle $-\alpha$ with the horizontal is also an axis of symmetry. Since there are only two axes of symmetry, it must be that $\ell = m$, which implies $\alpha = 90°$.

2.30. This is exactly **PROBLEM 2.3.3** in the text: if no line passes through exactly 2 of the points, it must be that one of the hypothesis in **PROBLEM 2.3.3** fails. Since we have finitely many points anyway, what must be failing is the hypothesis about collinearity. Thus, we must have that all the points are collinear.

2.31. Since the problem can be solved without knowing the radius, r, of the hole, the solution must be independent of r. We have to do a calculation, and we know that the final answer is independent of the value of r. So we might as well choose $r = 0$ (any other r would work, but it would make the computation much harder). In this case, the volume of the portion that remains is the volume of the original ball, and since the length of the hole is 6 inches, the diameter of this ball must be 6 inches. This implies that its volume is 36π. Thus, the volume of the portion that remains is 36π.

2.32. Suppose that the vertices of the square are $(0,0)$, $(1,0)$, $(0,1)$, $(1,1)$. We can assume that one vertex of the triangle, call it $B = (b,0)$, lies in the bottom side of the square, and the other two, call them $A = (0,a)$ and $C = (1,c)$, lie in the left and right side respectively, as in the following figure:

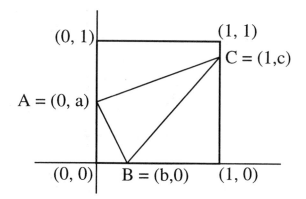

Then the area of the triangle is the area of the square minus the area of the trapezoid $AC(1,1)(0,1)$, the area of the triangle $\triangle BC(1,0)$ and the area of the triangle $\triangle AB(0,0)$. Thus, we have

$$\text{Area} = 1 - \left(\frac{[(1-a)+(1-c)]\cdot 1}{2} + \frac{c\cdot(1-b)}{2} + \frac{a\cdot b}{2} \right)$$
$$= \frac{a(1-b)+cb}{2}.$$

The above quantity attains its maximum when $b=1, c=1$ or when $b=0, a=1$. In any event, the area of the triangle will be $1/2$.

For the rectangle one can use the same strategy, now taking into account the lengths of the sides of the rectangle. The answer will also be $(1/2)\cdot$Area of rectangle$= 1/2$.

For the circle, note that we can always assume that one of the sides of the triangle is vertical. Place the circle in a coordinate system so that the segment with endpoints $(0,0)$ and $(1,0)$ is a diameter of the circle. Since we are assuming that one of the sides of the triangle is vertical, two of the vertices will have the same x-coordinate, call it a, let c be the x-coordinate of the other vertex, and assume $a \le c$ (note that, since we can reflect, this assumption does not imply any loss of generality). Then, considering the vertical side as the base of the triangle, the area of the triangle will be:
$$\text{Area} = \frac{2a(c-a)}{2} = a(c-a).$$

This quantity is greater with greater c, so we must have $c=1$. The maximum of what remains, i.e. $a(1-a)$, is attained when $a=1/2$ (note that the graph of $a(1-a)$ is a parabola opening down with the vertex at $a=1/2$). Thus the maximum is attained when $a=1/2$ and $c=1$, which corresponds to an equilateral triangle of area $1/4$.

2.33. Let $P = (p_1, p_2)$. Let $X = (x,0)$ be the point of intersection of the line with the x-axis, and let $Y = (0,y)$ be the point of intersection of the line with the y-axis. Finally, denote the point $(p_1,0)$ by R and the point $(0,p_2)$ by S. All this is depicted in the following figure:

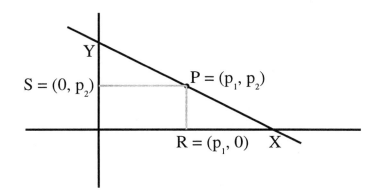

By similarity of the triangles $\triangle YSP$ and $\triangle RXP$, we have

$$\frac{(y - p_2)}{p_2} = \frac{p_1}{(x - p_1)}.$$

This implies

$$y = \frac{p_2 x}{(x - p_1)}.$$

Denote the area of the triangle by $A(x)$. Then we have $A(x) = (xy)/2$. Thus

$$A(x) = \frac{p_2}{2} \cdot \frac{x^2}{(x - p_1)}.$$

We want to find the value of x that gives the triangle with least area. Note that, from the inequality

$$0 \le (x - 2p_1)^2 = x^2 - 4p_1 \cdot (x - p_1),$$

we obtain

$$4p_1 \le \frac{x^2}{(x - p_1)},$$

which implies

$$A(x) \le 2p_1 p_2.$$

But $A(2p_1) = 2p_1 p_2$. Thus, the minimum area is attained when $x = 2p_1$ and $y = 2p_2$.

2.34. Divide the complex plane into 4 sectors S_1, S_2, S_3, S_4 centered in the axes:

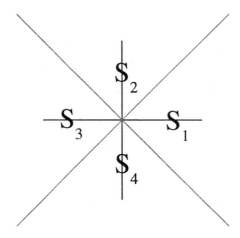

Let $T_j = \{a_k : a_k \in S_k\}$, $j = 1, 2, 3, 4$. If we have that

$$\sum_{a_\ell \in T_1} |a_\ell| \geq \frac{1}{10} \sum_1^m |a_j|,$$

then we are done, since

$$\left| \sum_{a_\ell \in T_1} a_\ell \right| \geq \frac{1}{\sqrt{2}} \sum_{a_\ell \in T_1} |a_j| \geq \frac{1}{10\sqrt{2}} \sum_1^m |a_j|.$$

Likewise, if we have that

$$\sum_{a_\ell \in T_2} |a_\ell| \geq \frac{1}{10} \sum_1^m |a_j|,$$

or

$$\sum_{a_\ell \in T_3} |a_\ell| \geq \frac{1}{10} \sum_1^m |a_j|,$$

then we are done.

If all of the above fail, then we must have that

$$\sum_{a_\ell \in T_1 \cup T_2 \cup T_3} |a_\ell| \leq \frac{3}{10} \sum_1^m |a_j|.$$

Hence,

$$\left| \sum_{a_\ell \in T_4} a_\ell \right| \geq \frac{1}{\sqrt{2}} \sum_{a_\ell \in T_4} |a_j|$$

$$= \geq \frac{1}{1\sqrt{2}} \left(\sum_1^m |a_j| - \frac{3}{10} \sum_1^m |a_j| \right)$$

$$= \frac{1}{2\sqrt{2}} \sum_1^m |a_j|.$$

This is what we wanted to prove. Note that the constant C is $1/(10\sqrt{2})$.

2.36. For three distinct points A, B, C in the plane, define the 'excess' of A, B, C (denoted $E(A, B, C)$) to be the minimum of the three quantities

$$(|A - B| + |B - C| - |C - A|), (|B - C| + |C - A| - |A - B|),$$
$$(|C - A| + |A - B| - |B - C|).$$

Notice that, by the triangle inequality, these quantities are all positive. Also, the excess a triple of points is 0 if and only if these points are collinear, and it is less than or equal than the distances between any two points of the triple. The reader should draw some pictures at this point to convince herself/himself of these important facts.

If we have a set of points in the plane that are at integer distances from each other, the excess of any triple of points in this set must be an integer (since in that case the three quantities involved in the definition of excess are integers).

Now take two points A and B in the set, and let $d \in \mathbf{Z}$ be the distance between them. Since the set consists of infinitely many points, it must contain points that are arbitrarily far from both A and B. Let X be a third point whose distance to both A and B is greater than d, and let $k := E(A, B, X) \leq d$ (remember that the excess of a triple of points is always less than or equal to the distance between any two points of the triple). Since

$$|B - X| + |X - A| - |A - B| > d + d - d > d \geq E(A, B, X),$$

we must have that either

$$E(A, B, X) = |A - B| + |B - X| - |X - A|$$

or

$$E(A, B, X) = |A - B| + |X - A| - |B - X|.$$

Reordering the last equation and using $E(A, B, X) = k$ and $|A - B| = d$ we have that, for X sufficiently far from A and B, X must satisfy either

$$d - k = |X - A| - |B - X|$$

or

$$d - k = |B - X| - |X - A|.$$

This means that all the points in the set that are sufficiently far from A and B lie either on one of the hyperbolae

$$d - k = \Big| |B - X| - |X - A| \Big|, \quad k = 0, 1, 2, \ldots, d - 1,$$

or on the line through A and B (this corresponds of the case $k = d$). (Recall that a hyperbola is the set of points in the plane satisfying that the difference of their distances to two fixed points called foci is a constant—this is exactly what the last equations say, taking A and B as the foci.)

Next we will show that each one of these hyperbolae contains at most finitely many points of the set. After performing a translation and maybe a rotation, we can depict one of these hyperbolae as follows:

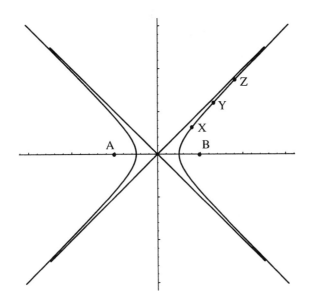

The lines through the origin in the figure above are called the asymptotes. The graph of the hyperbola gets closer and closer to these lines in each of the four branches. Now suppose that we have infinitely many points on one of the hyperbolae discussed above. Then one of the branches in that hyperbola must contain infinitely many points. Thus there must be points in that branch that are arbitrarily far from A and B. Let X, Y and Z be three points in that branch that are very far from A and B, as in the previous figure. Since the graph of that branch gets very close to the graph of a line when we are very far from A and B, the points X, Y and Z lie almost in a line (see the figure below). What happens to their excess? If we assume that X is farther from A and B than Y and Y is farther from A and B than Z (i.e. Y lies between X and Z in the hyperbola), then their excess will be

$$|Z - Y| + |Y - X| - |Z - X|.$$

But since all the points are very close to being collinear, this quantity can be made as small as we want if we take X, Y and Z far enough. On the other hand, since there are not really collinear, their excess must be nonzero. But the excess of any three points in the set defined in the statement of the exercise is an integer, and there are no nonzero integers that are arbitrarily small. This contradiction shows that only a finite number of points lie on each of the hyperbolae above

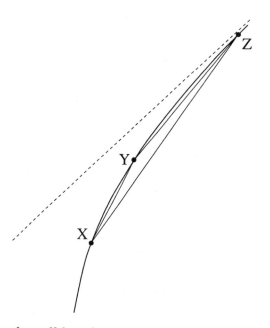

This means that all but finitely many points lie in the line through A and B. Next, we will show that no points can lie outside this line.

Assume that there is a point in the set that does not lie in this line. Let C be this point. Now, since there are infinitely many points in the line through A and B, we can find points V, W arbitrarily far from C. All this is depicted in the following figure:

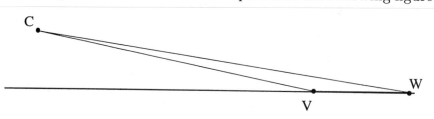

Now we reason as in the case of the hyperbolae. If W is farther from C than V, then the excess of C, V, W is

$$|C - V| + |V - W| - |C - W|$$

Now, when V and W are very far from C, the triangle $\triangle CVW$ is very flat, so that the sum of the distances $|C - V|$ and $|V - W|$ is very similar to the distance $|C - W|$. (This can be proved analytically; we encourage the reader to find an analytic proof of this fact.)

Thus, the excess of the triple C, V, W is nonzero but arbitrarily small. Since it also has to be an integer and there are no arbitrarily small nonzero integers, we arrive at a contradiction. Hence our assumption, namely that there is a point in the set that does not lie in the line through A and B is false. Therefore all the points are collinear.

2.39. Here is a way to cut a torus in 12 pieces with three planar cuts.

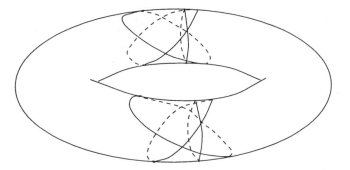

We do not know if this is the maximum number of pieces that one can get with three cuts.

2.41. The length of each of the approximations is exactly 2 (note that, in the nth approximation, we are going up n steps of length $1/n$ each and we are going left n steps of length $1/n$ each, so the total is 2). This suggests that the length of the diagonal is 2, but we know from the Pythagorean theorem that the length is $\sqrt{1^2 + 1^2} = \sqrt{2}$. This seems to be a contradiction. But it is not: the length of the limit of a sequence of sets need not be the limit of the lengths of the sets.

Note that two curves can be very close together but have completely different lengths. For example, look at the following picture. The saw-like grey curve is much longer, yet they are very close.

2.42. Here is an approximation to the circle. In the first figure, the approximation is composed of 8 segments of length 1. In the second figure, of 16 segments of length 1/2, etc. The limit of the length of the approximation is, therefore, 8.

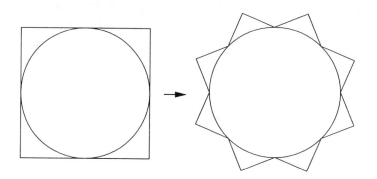

etc...

Chapter 3

Problems Involving Counting

3.1. Let the players be A and B. If player A receives one card, then there are n choices, namely all the ways to choose the card that player A will receive.

If player A receives 2 cards, then we have to count all possible ways to choose 2 cards out of a deck with n cards, i.e. $\binom{n}{2}$.

Continuing in this fashion, we have that if player A receives k cards, then the number of possible ways will be $\binom{n}{k}$.

Therefore, the total number of ways to deal the cards will be

$$\sum_{k=1}^{n-1} \binom{n}{k} = 2(2^{n-1} - 1).$$

This last equality can be proved as follows:

From the text we know that

$$\sum_{k=0}^{n} \binom{n}{k} = 2^n.$$

Therefore we have

$$
\begin{aligned}
\sum_{k=1}^{n-1} \binom{n}{k} &= \sum_{k=0}^{n} \binom{n}{k} - \binom{n}{0} - \binom{n}{n} \\
&= 2^n - 1 - 1 \\
&= 2(2^{n-1} - 1),
\end{aligned}
$$

as desired.

3.2. Let us number the women in the order they surreptitiously divided the pile of diamonds. Let N_0 denote the original number of diamonds, L the number of diamonds that each woman received at the last division and N_k the number of diamonds that woman k left. Then we have:

$$N_5 = 5L + 1$$

and

$$N_{k+1} = \frac{4}{5}(N_k - 1), \quad k = 0, 1, \ldots, 5.$$

Solving the recurrence relation we find that

$$
\begin{aligned}
N_0 &= \frac{5^6 L}{4^5} + \frac{5^5}{4^5} + \frac{5^4}{4^4} + \frac{5^3}{4^3} + \frac{5^2}{4^2} + \frac{5}{4} + 1 \\
&= \frac{5^6 L}{4^5} + \frac{5^6 - 4^6}{4^5}.
\end{aligned}
$$

Reordering, we obtain

$$4^5(N_0 + 4) = 5^6(L + 1).$$

Since 4 and 5 are relatively prime, we must have that $N_0 + 4$ divides 5^6 (note that $N_0 + 4$ cannot divide $L + 1$ since $N_0 + 4 > L + 1$). Since we are looking for the least N_0, we must have

$$N_0 = 5^6 - 4 = 15,621,$$

and

$$L = 4^6 - 1 = 1,023.$$

Thus, the original number of diamonds is $N_0 = 15,621$.

3.3. Let us study the parity of the first four numbers of each row. Let o stand for 'odd' and e stand for 'even'. The first four numbers of the third row are $\{1, e, o, e\}$. The first four numbers of the subsequent rows will be as follows:

$$
\begin{aligned}
3^{\text{rd}} \text{ row:} &\quad \{1, e, o, e, \ldots\} \\
4^{\text{th}} \text{ row:} &\quad \{1, o, e, o, \ldots\} \\
5^{\text{th}} \text{ row:} &\quad \{1, e, e, e, \ldots\} \\
6^{\text{th}} \text{ row:} &\quad \{1, o, o, e, \ldots\} \\
7^{\text{th}} \text{ row:} &\quad \{1, e, o, e, \ldots\}
\end{aligned}
$$

$$\vdots \qquad\qquad \vdots$$

After the 7th row, the pattern repeats. Since every row in the pattern above has an even number among the four first numbers and the pattern repeats, we must have that every row has at least one even number among its first four elements.

3.4. The only way that six coins can make exactly one dollar is if we have 3 quarters, 2 dimes and 1 nickel. To see this, note that we must have exactly 3 quarters, since with two quarters, the maximum amount we could have would only be 90 cents (using 4 dimes), and 4 quarters already make a dollar. For the remaining 25 cents we must have exactly 2 dimes, since with only one dime the maximum we can have is 20 cents (using 2 nickels), and with three dimes we have more than 25 cents.

Therefore, the probability that the lost coin was a dime is

$$\frac{\text{number of dimes}}{\text{number of coins}} = \frac{2}{6} = \frac{1}{3}.$$

3.5. The strategy we will use is the following: assuming that the couple has n children, find the probability that they have two sons (s) and one daughter (d). Then find n so that this probability is greater than one-half.

Given that the couple has n children,

$$
\begin{aligned}
\mathbf{Pr}\{2\text{ s, }1\text{ d}\} &= 1 - (\mathbf{Pr}\{\text{less than 2 s}\} + \mathbf{Pr}\{\text{less than 1 d}\}) \\
&= 1 - (\mathbf{Pr}\{0\text{ b}\} + \mathbf{Pr}\{1\text{ b}\} + \mathbf{Pr}\{0\text{ d}\}) \\
&= 1 - \left(\frac{1}{2^n} + \frac{n}{2^n} + \frac{1}{2^n}\right).
\end{aligned}
$$

We want to find the least n such that

$$1 - \left(\frac{1}{2^n} + \frac{n}{2^n} + \frac{1}{2^n}\right) > \frac{1}{2},$$

or, reordering,

$$2^{n-1} > n + 2.$$

By simple inspection we find that $n = 4$.

3.6. The strategy is to find the conditional probabilities that the next fly survives given that the spider has already nabbed k flies, where $k = 0, 1, 2$ or 3. Then we use the formula

$$\mathbf{Pr}\{\text{next fly survives}\}$$

$$= \sum_{k=0}^{3} \mathbf{Pr}\{k \text{ were nabbed}\} \cdot \mathbf{Pr}\{\text{next survives } | k \text{ were nabbed}\}.$$

The probability that exactly k were nabbed can be calculated as follows: there are $\binom{5}{k}$ possible ways in which k flies could have been nabbed. The total number of outcomes is $\binom{5}{0} + \binom{5}{1} + \binom{5}{2} + \binom{5}{3} = 26$, since the cases where 4 or 5 flies were nabbed are not possible (the spider eats at most 3 flies). Thus we have

$$\mathbf{Pr}\{k \text{ were nabbed}\} = \frac{\binom{5}{k}}{26}.$$

On the other hand,

$$\mathbf{Pr}\{\text{next survives } | k \text{ were nabbed}\} = \begin{cases} 1/2 & \text{if } k < 3 \\ 1 & \text{if } k = 3 \end{cases}$$

Thus,

$$\begin{aligned}
\mathbf{Pr}\{\text{next fly survives}\} &= \frac{1}{2}\left[\frac{1}{26}\binom{5}{0} + \frac{1}{26}\binom{5}{1} + \frac{1}{26}\binom{5}{2}\right] + \frac{1}{26}\binom{5}{3} \\
&= \frac{1}{52}(1 + 5 + 10 + 20) \\
&= \frac{36}{52} \\
&= \frac{9}{13}.
\end{aligned}$$

3.7. This problem is quite hard, and requires some more advanced notions of probability. We are receiving one card out of a deck of 52 for each purchase of a pack of baseball cards. Let us do a more general case and assume that for each purchase we receive one card out of a deck of n cards, which we will number from 1 to n. Let X_n be the number of purchases made when we complete the whole deck of n cards. We want to find

$$E[X] = 1 \cdot \mathbf{Pr}\{X = 1\} + 2 \cdot \mathbf{Pr}\{X = 2\} + \cdots + m \cdot \mathbf{Pr}\{X = m\} + \cdots.$$

Let us first find the probability that $X_n \geq k$. For $k \leq n$, this probability is always 1. In general, it is quite complicated, but it can be found as follows:

$$\mathbf{Pr}\{X_n \geq k\} = \mathbf{Pr}\{\text{some card is missing after } k - 1 \text{ purchases}\}.$$

Now, the probability that some card is missing is the probability of the union of the events

$$A_i^k = \{\text{card } i \text{ is missing after } k - 1 \text{ purchases}\}, \ 1 \leq i \leq n.$$

The probability of the union of n events can be found using the following formula. The proof of this formula is not hard using induction, and is left as an exercise for the reader. It is best to work out a few easy cases (say $n = 1, 2, 3$) to see that the formula makes sense (see also Exercise 19 in this chapter).

$$\mathbf{Pr}\{A_1^k \cup A_2^k \cup A_3^k \cup \cdots \cup A_n^k\}$$
$$= \sum_{\ell=1}^{n} \mathbf{Pr}\{A_i^k\} - \sum_{i<j} \mathbf{Pr}\{A_i^k \cap A_j^k\} + \sum_{i_1<i_2<i_3} \mathbf{Pr}\{A_{i_1}^k \cap A_{i_2}^k \cap A_{i_3}^k\}$$
$$+ \cdots + (-1)^{d+1} \sum_{i_1<i_2<\cdots<i_d} \mathbf{Pr}\{A_{i_1}^k \cap A_{i_2}^k \cap \cdots \cap A_{i_d}^k\}$$
$$+ \cdots + (-1)^{n+1} \mathbf{Pr}\{A_1^k \cap A_2^k \cap A_3^k \cap \cdots \cap A_n^k\}.$$

Now, $\mathbf{Pr}\{A_i\}$ is the same for every i (the probability that card 1 is missing is the same as the probability that any other card is missing). The probability that card i is missing after $k - 1$ purchases is

$$\left(\frac{n-1}{n}\right)^{k-1}.$$

The probability that two cards are missing (i.e. $\mathbf{Pr}\{A_i^k \cap A_j^k\}$) is also independent of i and j, and equals

$$\left(\frac{n-2}{n}\right)^{k-1}.$$

In general, the probability that d cards are missing, (in other words, $\mathbf{Pr}\{A_{i_1}^k \cap A_{i_2}^k \cap \cdots \cap A_{i_d}^k\}$), is also independent of $i_1, \ldots i_d$, and equals

$$\left(\frac{n-d}{n}\right)^{k-1}.$$

Thus, applying the formula above we obtain

$$\mathbf{Pr}\{X_n \geq k\}$$
$$= \sum_{\ell=1}^{n}\left(\frac{n-1}{n}\right)^{k-1} - \sum_{i<j}\left(\frac{n-2}{n}\right)^{k-1} + \sum_{i_1<i_2<i_3}\left(\frac{n-3}{n}\right)^{k-1}$$
$$+ \cdots + (-1)^{d+1}\sum_{i_1<i_2<\cdots<i_d}\left(\frac{n-d}{n}\right)^{k-1}$$
$$+ \cdots + (-1)^{n+1}\left(\frac{n-n}{n}\right)^{k-1}.$$

The summation in the third line of the equation above has $\binom{n}{d}$ terms. Using this fact and skipping the last term (which is 0), we can rewrite the last expression as

$$\mathbf{Pr}\{X_n \geq k\}$$
$$= \binom{n}{1}\left(\frac{n-1}{n}\right)^{k-1} - \binom{n}{2}\left(\frac{n-2}{n}\right)^{k-1} + \binom{n}{3}\left(\frac{n-3}{n}\right)^{k-1}$$
$$+ \cdots + (-1)^{d+1}\binom{n}{d}\left(\frac{n-d}{n}\right)^{k-1}$$
$$+ \cdots + (-1)^{n}\binom{n}{n-1}\left(\frac{1}{n}\right)^{k-1}.$$

This can be rewritten in a more compact form as

$$\mathbf{Pr}\{X_n \geq k\} = \sum_{d=1}^{n-1}(-1)^{d+1}\binom{n}{d}\left(\frac{n-d}{n}\right)^{k-1}.$$

Once we have found $\mathbf{Pr}\{X_n \geq k\}$, we could find $\mathbf{Pr}\{X_n = k\}$ and use the formula for $\mathrm{E}[X_n]$ above. Instead we will use the following: We have

$$\mathrm{E}[X_n] = \sum_{k=1}^{\infty} i \cdot \mathbf{Pr}\{X_n = k\}.$$

Since multiplying by i is the same as adding i times, we can rewrite the last expression as

$$\mathrm{E}[X_n] = \sum_{i=k}^{\infty} \sum_{j=1}^{k} \mathbf{Pr}\{X_n = k\}.$$

Reversing the order of summation we obtain

$$\mathrm{E}[X_n] = \sum_{j=k}^{\infty} \sum_{k=j}^{\infty} \mathbf{Pr}\{X_n = k\}.$$

But we also have

$$\sum_{k=j}^{\infty} \mathbf{Pr}\{X_n = k\} = \mathbf{Pr}\{X_n \geq j\}.$$

Thus, we obtain the formula

$$\mathrm{E}[X_n] = \sum_{j=1}^{\infty} \mathbf{Pr}\{X_n \geq j\}.$$

Therefore, in our case we have (recall that $\mathbf{Pr}\{X \geq k\} = 1$ for $k \leq n$)

$$\mathrm{E}[X_n] = \sum_{j=1}^{n} 1 + \sum_{j=n+1}^{\infty} \sum_{d=1}^{n-1} (-1)^{d+1} \binom{n}{d} \left(\frac{n-d}{n}\right)^{j-1}.$$

Simplify the first sum and reverse the order of summation in the second:

$$\mathrm{E}[X_n] = n + \sum_{d=1}^{n-1} (-1)^{d+1} \binom{n}{d} \sum_{j=n+1}^{\infty} \left(\frac{n-d}{n}\right)^{j-1}.$$

Using the formula for a geometric series, we get

$$E[X_n] = n + \sum_{d=1}^{n-1} (-1)^{d+1} \binom{n}{d} \frac{n}{d} \left(\frac{n-d}{n} \right)^n.$$

This is actually the final solution. The only thing that remains is to substitute $n = 52$. Calculating this sum by hand is very tedious. Fortunately, we can use a computer to do the job. For $n = 52$ and using Mathematica we obtained $E[X_{52}] = 235.978$.

3.8. If Bluesky is the first thrower, the probability that he wins is:

Pr{Bluesky wins}
= **Pr**{1st throw hits}
+ **Pr**{3rd throw hits|1st & 2nd missed} · **Pr**{1st & 2nd missed}
+ **Pr**{5rd throw hits|1st to 4th missed} · **Pr**{1st to 4th missed}
+ · · ·

Now,

$$\textbf{Pr}\{n^{\text{th}} \text{ throw hits}|1^{\text{st}} \text{ to } (n-1)^{\text{th}} \text{ throws were missed}\} = \frac{1}{2},$$

since the probability of hitting the target is $1/2$ no matter what happened before.

$$\textbf{Pr}\{ 1^{\text{st}}, 2^{\text{nd}}, \ldots (n-1)^{\text{th}} \text{ throws were missed}\} = \frac{1}{2^{n-1}}.$$

Therefore we have

$$
\begin{aligned}
\textbf{Pr}\{\text{Bluesky wins}\} &= \frac{1}{2} + \frac{1}{2} \cdot \frac{1}{2^2} + \frac{1}{2} \cdot \frac{1}{2^4} + \frac{1}{2} \cdot \frac{1}{2^6} + \cdots \\
&= \frac{1}{2} \sum_{n=0}^{\infty} \left(\frac{1}{4} \right)^n \\
&= \frac{1}{2} \cdot \frac{1}{1 - 1/4} \\
&= \frac{2}{3}.
\end{aligned}
$$

3.9. If all the points lie in the same half disc, then there is an angle formed by the rays through two of the points that contains the third point, and the measure of this angle is less than or equal to 180°. If we replace the middle point with its antipodal point then the points do not lie in the same half disc any more.

In other words, substituting the middle point with its antipodal gives a function from the set {configurations lying in the same half disc} to the set {configurations not lying in the same half disc}. This function is 3 to 1: each configuration of points not lying in the same half disk has three preimages under our function, as in the picture below:

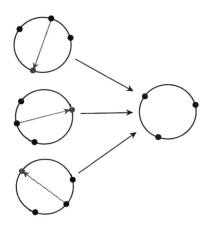

Thus, the sets {configurations lying in the same half disc} and {configurations not lying in the same half disc} are in 3 to 1 correspondence. Thus, the probability that the points do not lie in the same half disk is 1/4.

3.11. Although this problem seems similar to the previous two, it is actually much easier. We are assuming that the square has been divided beforehand, either diagonally, horizontally or in any other way. Then the probability that each of the points lies in a given half square is 1/2. Therefore, the probability that all three points lie in a given half square is

$$\frac{1}{2} \cdot \frac{1}{2} \cdot \frac{1}{2} = \frac{1}{8}.$$

Since there are two half squares, the probability that all three points lie in the same half square is $2 \cdot 1/8 = 1/4$.

3.12. The 4 numbers can be drawn in $4! = 24$ different orderings, and each ordering is equally likely to happen. Therefore the probability that they are drawn in increasing order is $1/24$.

3.13. In the previous exercise we only took into account the fact that there was an ordering, not the values of the numbers involved. In this exercise, even though we do not know what the numbers are or how they are sequenced, we do know that there is an ordering, and that is the only thing that matters. Therefore the answer is the same as in the previous exercise..

3.14. When we draw the first one, we have $5/9$ probability of selecting a red one. Once we have selected the first red, we have $4/8$ probability of selecting another red one (note that only 4 red ones and 4 blue ones are still in the bag). For the next one we have a $3/7$ probability, for the fourth one $2/6$ and for the last one $1/5$. Thus the probability of selecting five red marbles is

$$\frac{5}{9} \cdot \frac{4}{8} \cdot \frac{3}{7} \cdot \frac{2}{6} \cdot \frac{1}{5} = \frac{1}{126}.$$

3.15. *Step 1:* Fill the 8 quart container. Then fill up the 5 quart container using the contents of the 8 quart container. Then 3 quarts remain in the 8 quart container.

Step 2: Empty the 5 quart container and pour the 3 quarts from the 8 quart container into it. Then fill up the 8 quart container again and top-off the 5 quart container using the contents of the 8 quart container. Since the 5 quart container already had 3 quarts of water, 6 quarts still remain in the 8 quart container.

Step 3: Empty the 5 quart container and fill it up using the contents of the 8 quart container again. Since we had 6 quarts in the 8 quart container, only 1 quart remains in the 8 quart container, as desired.

3.16. We just have to count the number of digits used:

From 1 to 9	we wrote	$9 \cdot 1 = 9$ digits.
From 10 to 99	we wrote	$90 \cdot 2 = 180$ digits.
From 100 to 999	we wrote	$900 \cdot 3 = 2,700$ digits.
From 1,000 to 9,999	we wrote	$9,000 \cdot 4 = 36,000$ digits.
From 10,000 to 12,199	we wrote	$2,200 \cdot 5 = 11,000$ digits.
From 12,200 to 12,221	we wrote	$22 \cdot 5 = 110$ digits.
	TOTAL	49,999 digits.

Therefore, the $50,000^{\text{th}}$ number we wrote is the 1 of 12,222.

3.17. Proceed as follows:

From 1 to 9	there are	$9 \cdot 1 = 9$ digits.
From 10 to 99	there are	$90 \cdot 2 = 180$ digits.
From 100 to 750	there are	$651 \cdot 3 = 1,953$ digits.
	TOTAL	2,141 digits.

It will require 2,141 digits.

3.18. Let the lengths of the sides be a, b, s. We know that $0 \leq s \leq a+b$, $0 \leq a \leq s+b$ and $0 \leq b \leq a+s$. We want to find

$$\mathbf{Pr}\left\{0 \leq s \leq \frac{a+b}{2}\right\}.$$

Assume that $a \leq b$ (note that this happens with probability 1/2). Then we are asking

$$\mathbf{Pr}\left\{b - a \leq s \leq \frac{a+b}{2}\right\}.$$

Note that, in general, $s \in [0, a+b]$, and we assume that it takes any value in this interval with equal probability. Then the probability above is just the ratio of the lengths of $[b - a, (a+b)/2]$ and $[0, a+b]$, i.e. it equals

$$\frac{\frac{a+b}{2} - (b-a)}{a+b} = \frac{3a - b}{2(a+b)}.$$

Thus we have

$$\mathbf{Pr}\left\{0 \le s \le \frac{a+b}{2}\right\}$$

$$= \mathbf{Pr}\left\{b-a \le s \le \frac{a+b}{2}\,\bigg|\,a \le b\right\} \cdot \mathbf{Pr}\{a \le b\}$$

$$+\mathbf{Pr}\left\{a-b \le s \le \frac{a+b}{2}\,\bigg|\,a \le b\right\} \cdot \mathbf{Pr}\{b \le a\}$$

$$= \frac{3a-b}{2(a+b)} \cdot \frac{1}{2} + \frac{3b-a}{2(a+b)} \cdot \frac{1}{2}$$

$$= \frac{1}{2}.$$

3.19. An easy way to do this kind of problems is by means of a Venn diagram. There is also a formula—which is not hard to prove after looking at a Venn diagram—that reads as follows: If A, B, C are finite sets,

$$|A \cup B \cup C| = |A|+|B|+|C|-|A \cap B|-|B \cap C|-|C \cap A|+|A \cap B \cap C|,$$

where $|\cdot|$ means 'number of elements'. In our case, let A, B and C stand for 'Algebra', 'Biology' and 'Chemistry' respectively. We have

$$\#\{\text{Students who failed any exam}\} = 12+5+8-2-6-3+1 = 15.$$

We can see this in the following Venn diagram:

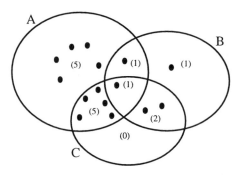

3.20. Before knowing that we have rolled at least a 9, the probability of rolling a 9 was 1/9, of rolling a 10 was 1/12, of rolling an 11 was 1/18, and of rolling a 12 was 1/36. With the additional information, the probabilities change: they now have to add up to 1, since we know *for sure* that we have rolled at least a 9. But the proportions between the probabilities remain unchanged. Therefore, since we want them to add up to 1, we have to divide each of the probabilities by $1/9 + 1/12 + 1/18 + 1/36 = 5/18$. So we have:

$$\mathbf{Pr}\{9 \mid \text{score} \geq 9\} = \frac{1}{9} \cdot \frac{18}{5} = \frac{2}{5}$$

$$\mathbf{Pr}\{10 \mid \text{score} \geq 9\} = \frac{1}{12} \cdot \frac{18}{5} = \frac{3}{10}$$

$$\mathbf{Pr}\{11 \mid \text{score} \geq 9\} = \frac{1}{18} \cdot \frac{18}{5} = \frac{1}{5}$$

$$\mathbf{Pr}\{12 \mid \text{score} \geq 9\} = \frac{1}{36} \cdot \frac{18}{5} = \frac{1}{10}.$$

Thus, the conditional probability of having rolled an 11 is $1/5$ and the probability of having rolled at least 11 is $\mathbf{Pr}\{11\} + \mathbf{Pr}\{12\} = 3/10$.

3.21. It is actually true that most people come from larger than average families. This can be proved as follows: let f_i be the number of families with i children, and let c_i be the number of children that come from families of size i. We must have $c_i = if_i$. The average family size is the sum of the number of families times their size divided by the total number of families, i.e.

$$\text{Av}_f = \frac{\sum\limits_{i=1}^{K} if_i}{\sum\limits_{i=1}^{K} f_i},$$

where K the maximum number of children found in any family.

On the other hand, if we pick a person, how big will her family be, on average? This is a different question. We have to add the number of people times the size of each person's family, and then divide by the total number of people, i.e.

$$\mathrm{Av}_p = \frac{\displaystyle\sum_{i=1}^{K} ic_i}{\displaystyle\sum_{i=1}^{K} c_i} = \frac{\displaystyle\sum_{i=1}^{K} i^2 f_i}{\displaystyle\sum_{i=1}^{K} if_i}.$$

We will show that $\mathrm{Av}_f \leq \mathrm{Av}_c$, thus showing that if we pick a person in the street then, on average, she will come from a larger than average family.

We need to show that

$$\frac{\displaystyle\sum_{i=1}^{K} if_i}{\displaystyle\sum_{i=1}^{K} f_i} \leq \frac{\displaystyle\sum_{i=1}^{K} i^2 f_i}{\displaystyle\sum_{i=1}^{K} if_i}$$

or, equivalently,

$$\left(\sum_{i=1}^{K} if_i\right)\left(\sum_{i=1}^{K} if_i\right) \leq \left(\sum_{i=1}^{K} i^2 f_i\right)\left(\sum_{i=1}^{K} f_i\right).$$

Multiplying this out,

$$\sum_{i=1}^{K} ij f_i f_j \leq \sum_{i=1}^{K} i^2 f_i f_j.$$

We can split the sums to obtain

$$\sum_{1\leq i\leq j\leq K} ij f_i f_j + \sum_{1\leq j< i\leq K} ij f_i f_j \leq \sum_{1\leq i\leq j\leq K} i^2 f_i f_j + \sum_{1\leq j< i\leq K} i^2 f_i f_j.$$

Interchanging i and j in the last summation of both sides (this is just a change of variables), we get

$$\sum_{1\leq i\leq j\leq K} ij f_i f_j + \sum_{1\leq i< j\leq K} ji f_j f_i \leq \sum_{1\leq i\leq j\leq K} i^2 f_i f_j + \sum_{1\leq i< j\leq K} j^2 f_j f_i.$$

Rearranging, we have

$$\sum_{1\leq i<j\leq K} ijf_if_j + \sum_{1\leq i<j\leq K} jif_jf_i + \sum_{1\leq i\leq K} i^2f_i^2$$

$$\leq \sum_{1\leq i<j\leq K} i^2f_if_j + \sum_{1\leq i<j\leq K} j^2f_jf_i + \sum_{1\leq i\leq K} i^2f_i^2.$$

We can subtract $\sum_{1\leq i\leq K} i^2f_i^2$ from both sides to obtain

$$\sum_{1\leq i<j\leq K} ijf_if_j + \sum_{1\leq i<j\leq K} jif_jf_i \leq \sum_{1\leq i<j\leq K} i^2f_if_j + \sum_{1\leq i<j\leq K} j^2f_jf_i.$$

Now, both terms in the left hand side are equal, and in the right hand side we can factor f_jf_i. We obtain

$$\sum_{1\leq i<j\leq K} 2ijf_if_j \leq \sum_{1\leq i<j\leq K} (i^2+j^2)f_if_j.$$

Finally, note that since $0 \leq (i-j)^2 = i^2 + j^2 - 2ij$, we have $2ij \leq i^2 + j^2$. Thus, the coefficients of f_if_j in the left hand side are always less than or equal to the corresponding coefficients in the right hand side. Thus, the inequality is true, proving that most people come from larger than average families.

3.22. There is an algorithm that always gives a final satisfactory arrangement. First, let us denote the 5 women by A, B, C, D, E and the 5 men by α, β, γ, δ, ϵ. We can express the rankings of men by women and the rankings of women by men in a matrix. For example, in the matrix

	A	B	C	D	E
α	$(1,2)$	$(2,4)$	$(3,3)$	$(4,5)$	$(5,1)$
β	$(1,1)$	$(3,3)$	$(4,1)$	$(5,1)$	$(2,4)$
γ	$(4,3)$	$(3,1)$	$(2,2)$	$(1,3)$	$(5,5)$
δ	$(3,5)$	$(2,2)$	$(1,4)$	$(5,4)$	$(4,3)$
ϵ	$(3,4)$	$(2,5)$	$(1,5)$	$(4,2)$	$(2,2)$

The first number in the pair gives the rating that the man in that row gives to the woman in that column, and the second digit gives the rating that the woman in that column gives to the man in that row. We encourage the reader to write down some examples and try to find a satisfactory arrangement in each case.

The algorithm cited above works as follows:

Step 1: Let the women propose to the men who are their first choices. Each man will provisionally accept only the proposal that is higest in his ranking. So at the end we have a number of provisional couples, some men that have not been proposed, and some women that got rejected. For the example given by the matrix above, at this stage we will have the couples (A, β), (B, γ), (E, α); δ and ϵ have not been proposed yet and C and D were rejected.

Step 2: Let the women that were rejected in the previous step propose to the next men in their lists. Each man will pick his favorite from the group consisting of the new proposers and his provisional match from the previous step. As before, at the end we have a number of provisional couples, some men that have not been proposed, and some women that got rejected. For the example given by the matrix above, at this stage C will propose to γ (γ is the second best for C) and D will propose to ϵ. Then γ will break up with B (his match from the previous step) and take C as his provisional couple. Since ϵ had not been proposed before, he will provisionally accept D's proposal. Therefore we have the matches (A, β), (C, γ), (E, α), (D, ϵ), δ has not been proposed yet and B has been rejected.

Continue in this fashion. Once all men have been proposed and all women have been accepted, stop. For the previous example, it goes as follows:

Step 3: Only B remains, so she proposes to her second favorite (note that so far she has only proposed to her first choice), which is δ. Since δ had not been proposed before, he simply accepts the offer and pairs up with B. Now all the men have been proposed and all the women have a match, so we stop here. The final arrangement is

$$(A, \beta), \quad (B, \delta), \quad (C, \gamma), \quad (D, \epsilon), \quad (E, \alpha).$$

Note that in this arrangement, all the women got their first or second choices, whereas two of the men ended up marrying their fourth and fifth choices. In fact, if we reverse the process (this is, let the men propose instead of the women), the arrangements will be more favorable for the men than for the women.

The reason why the configurations obtained with this algorithm are satisfactory in the sense stated in the exercise is the following: suppose that John and Mary are in the club and they are not married to each other. If Mary prefers John to her own husband, he must have proposed to John at some point (since she proposed to everybody that had higher ranking than her husband before she married him). Hence she was rejected by John at some point, which implies that John prefers his wife to Mary. Hence no rearrangements will make the situation more satisfactory for everybody in the group.

We also refer the interested reader to the article *College Admissions and the Stability of Marriage,* by D. Gale and L. S. Shapley, published in the American Mathematical Monthly, # 69 (1962), pp. 9-15.

For the interested reader's amusement, we offer the following exercise that comes from the article cited above. Apply the algorithm to the matching problem consisting of four couples with rankings matrix

$$
\begin{array}{ccccc}
 & A & B & C & D \\
\alpha & (1,3) & (2,2) & (3,1) & (4,3) \\
\beta & (1,4) & (2,3) & (3,2) & (4,4) \\
\gamma & (3,1) & (1,4) & (2,3) & (4,2) \\
\delta & (2,2) & (3,1) & (1,4) & (4,1)
\end{array}
$$

You will need to do 10 steps before arriving at the final configuration.

3.23. In how many ways can we distribute l grapes into k glasses, subject to the given rules? Well, by definition, exactly $\binom{k}{l}$. Thus, if we count the number of subsets with 0 elements, plus the number of subsets with 1 element, plus the number of subsets with

2 elements, ..., plus the number of subsets with k elements, we have the sum

$$\binom{k}{0} + \binom{k}{1} + \binom{k}{2} + \cdots + \binom{k}{k}.$$

By the binomial theorem, this equals 2^k.

3.24. We assume that a couple has the same likelihood of conceiving a boy or a girl. Write B for boy and G for girl. Then the possible combinations we have are BB, BG, GB, GG. The case BB is ruled out, since they have at least a girl. Thus, the combinations that remain are BG, GB, GG, each with probability 1/3. Now, we see that in two of these combinations there is a boy. Therefore, the probability that her sibling is a boy is 2/3.

3.25. The probability that the other side is also red is the probability that we chose the card with both sides red at the beginning. The probability that we chose the card with both sides red at the beginning given that one of the sides of our card is red is 2/3: there are 2 favorable cases (each of the red faces of the all-red card) and 3 possible cases (the 3 faces we have not seen yet). Thus the probability that the other side is also red is 2/3.

3.26. The probability of rolling at least a five in each roll is $1/3$. Let us call this a 'success'. Thus the probability of success is $1/3$. The probability of rolling at least a five exactly five times is

$$(\mathbf{Pr}\{\text{success}\})^5 \cdot \mathbf{Pr}\{\text{failure}\} \cdot \#\{\text{ways to arrange success/failure}\}$$

$$= \left(\frac{1}{3}\right)^5 \cdot \frac{2}{3} \cdot \binom{6}{5}.$$

The probability of rolling at least a five exactly six times is

$$(\mathbf{Pr}\{\text{success}\})^6 = \left(\frac{1}{3}\right)^6.$$

The probability of rolling at least a five at least five times is the sum of the previous probabilities:

$$\mathbf{Pr}\{\text{At least a five at least five times}\} = \left(\frac{1}{3}\right)^5 \cdot \frac{2}{3} \cdot \binom{6}{5} + \left(\frac{1}{3}\right)^6$$

$$= \frac{13}{3^6}.$$

In probability theory, this is called a Bernoulli trial.

3.27. The number 30^4 can be decomposed as $2^4 3^4 5^4$. Any divisor of 30^4 can be decomposed uniquely as $2^a 3^b 5^c$, with a, b, c nonnegative integers not exceeding 4. Thus the number of divisors is exactly the number of ordered triples (a, b, c), where a, b and c are integers between 0 and 4 inclusive. This number is exactly $5 \cdot 5 \cdot 5 = 125$.

3.28. Assume the contrary, i.e. that we can partition $\{1, 2, 3, 4, 5\}$ into two disjoint sets such that none of these two subsets contain two numbers and their difference. Let us denote the subsets by A and B.

First note that 1 and 2 cannot be in the same subset since $2 - 1 = 1$. Let us say that 1 is in A and 2 is in B. Now note that 2 and 4 cannot be in the same subset either, since $4 - 2 = 2$. Thus, 4 must be in A. Now, 3 cannot be in A because 4 and 1 are in A

and $4 - 1 = 3$. Therefore, so far we must have that 1 and 4 are in A and 2 and 3 are in B.

Finally, 5 cannot be in A because 1 and 4 are in A and $5 - 4 = 1$. Therefore 5 must be in B. But 5 cannot be in B either, because 2 and 3 are in B and $5 - 3 = 2$. Thus we arrive at a contradiction, which implies that our original assumption (being able to partition $\{1, 2, 3, 4, 5\}$ into two disjoint sets such that none of these two subsets contain two numbers and their difference) is false.

3.29. We just have to count the number of digits used:

From 1 to 9	there are	$9 \cdot 1 = 9$ digits.
From 10 to 99	there are	$90 \cdot 2 = 180$ digits.
From 100 to 599	there are	$500 \cdot 3 = 1,500$ digits.
From 600 to 659	there are	$60 \cdot 3 = 180$ digits.
From 660 to 666	there are	$7 \cdot 3 = 21$ digits.
	TOTAL	$1,890$ digits.

Therefore, the book has 666 pages.

3.30. The general pattern is

$$(n^2 + 1) + (n^2 + 2) + \cdots + (n + 1)^2 = n^3 + (n + 1)^3.$$

To prove this formula, we will use the summation formula proved in the text:

$$\sum_{k=M+1}^{N} k = (M+1) + (M+2) + \cdots + N = \frac{(M + N + 1)(N - M)}{2}.$$

$$
\begin{aligned}
(n^2 + 1) &+ (n^2 + 2) + \cdots + (n + 1)^2 \\
&= \frac{[n^2 + 1 + (n + 1)^2] \cdot [(n + 1)^2 - n^2]}{2} \\
&= 1 + 3n + 3n^2 + 2n^3 \\
&= n^3 + (n + 1)^3.
\end{aligned}
$$

3.31. The general pattern is

$$1^2 - 2^2 + 3^2 - 4^2 + \cdots + (-1)^{n+1} n^2 = (-1)^{n+1} \cdot (1 + 2 + 3 + \cdots + n).$$

For n even, the left hand side can be written as

$$
\begin{aligned}
(1^2 - 2^2) &+ (3^2 - 4^2) + \cdots + ((n-1)^2 - n^2) \\
&= (1-2)(1+2) + (3-4)(3+4) + \cdots \\
&\qquad\qquad\qquad\qquad + ((n-1) - n)((n-1) + n) \\
&= -(1+2) - (3+4) - \cdots - ((n-1) + n) \\
&= -(1 + 2 + 3 + \cdots + n), \qquad \text{which is the desired formula.}
\end{aligned}
$$

For n odd, the left hand side can be written as

$$
\begin{aligned}
(1^2 - 2^2) &+ (3^2 - 4^2) + \cdots + ((n-2)^2 - (n-1)^2) + n^2 \\
&= (1-2)(1+2) + (3-4)(3+4) + \cdots \\
&\qquad\qquad + ((n-2) - (n-1))((n-2) + (n-1)) + n^2 \\
&= -(1+2) - (3+4) - \cdots - ((n-2) + (n-1)) + n^2 \\
&= -(1 + 2 + 3 + \cdots + (n-1)) + 2(1 + 2 + 3 + \cdots n) - n \\
&= 1 + 2 + 3 + \cdots n, \qquad \text{which is the desired formula.}
\end{aligned}
$$

Note that we have used the fact that, using the formula stated in the solution to the previous exercise, we have

$$
\begin{aligned}
2(1 + 2 + 3 + \cdots n) - n &= 2 \cdot \frac{(n+1)\,n}{2} - n \\
&= n^2 + n - n \\
&= n^2.
\end{aligned}
$$

3.32. If he wants to put a different number of dimes in each pocket, he must have at least $0 + 1 + 2 + \cdots + 9 = 45$ dimes. Since he only has 44 dimes, he cannot do it.

3.33. The pattern is

$$(n^2 - n + 1) + (n^2 - n + 3) + (n^2 - n + 5) + \cdots + (n^2 - n + (2n-1)) = n^3.$$

Note that the left hand side has n summands. Thus it can be written as

$$n \cdot n^2 - n \cdot n + (1 + 3 + 5 + \cdots + 2n - 1).$$

Now, $1 + 3 + 5 + \cdots + 2n - 1$ equals

$$
\begin{aligned}
&1 + 3 + 5 + \cdots + 2n - 1 \\
&= (2 \cdot 1 - 1) + (2 \cdot 2 - 1) + (2 \cdot 3 - 1) + \cdots + (2 \cdot n - 1) \\
&= 2(1 + 2 + 3 + \cdots + n) - \underbrace{(1 + 1 + 1 + \cdots + 1)}_{n\text{-summands}} \\
&= 2 \cdot \frac{n^2 + n}{2} - n \\
&= n^2.
\end{aligned}
$$

Thus we have that the expression above equals:

$$n \cdot n^2 - n \cdot n + (1 + 3 + 5 + \cdots + 2n - 1) = n^3 + n^2 - n^2 = n^3,$$

which is the desired result.

3.34. We have to solve the following equation:

$$50h + 25q + 10d + 5n + p = 50,$$

where h is the number of half dollars, q is the number of quarters, d is the number of dimes, n is the number of nickels, and p is the number of pennies. Note that, since all the other terms are multiples of 5, p must also be a multiple of 5.

First, $h = 0$ or 1. For $h = 1$ there is only one way. So let us take $h = 0$. We have to solve the equation

$$25q + 10d + 5n + p = 50.$$

It is clear that q must be 0, 1 or 2. For $q = 2$ there is only one case. So far we have counted a total of 2 ways to compose 50 cents. Let us take $q = 1$. So now we have to solve the equation

$$10d + 5n + p = 25.$$

In this equation, d can be 0, 1 or 2.

For $d = 0$, we have to solve $5n + p = 25$, which has 6 solutions: $n = 0, 1, 2, 3, 4$ or 5. Thus, so far we have a total of $6 + 1 + 1 = 8$.

For $d = 1$, we have to solve $5n + p = 15$, which has 4 solutions: $n = 0, 1, 2$ or 3. Thus, so far we have a total of $8 + 4 = 12$.

For $d = 2$, we have to solve $5n + p = 5$, which has 2 solutions: $n = 0$ or 1. Thus, so far we have a total of $12 + 2 = 14$.

This exhausts all the cases for $q = 1$. Let us assume now that $q = 0$. We have to solve the equation

$$10d + 5n + p = 50.$$

The number d can be 0, 1, 2, 3, 4 or 5. Let us count the number of solutions for each case:

$d = 0$ gives 11 solutions, namely $n = 0, 1, 2, \ldots 10$.
$d = 1$ gives 9 solutions, namely $n = 0, 1, 2, \ldots 8$.
$d = 2$ gives 7 solutions, namely $n = 0, 1, 2, \ldots 6$.
$d = 3$ gives 5 solutions, namely $n = 0, 1, 2, 3, 4$.
$d = 4$ gives 3 solutions, namely $n = 0, 1, 2$.
$d = 5$ gives 1 solution, namely $n = 0$.

Thus, the total number of solutions for making 50 cents is

$$14 + 11 + 9 + 7 + 5 + 3 + 1 = 50.$$

To compose 25 cents we can proceed similarly. We have to solve the equation

$$25q + 10d + 5n + p = 25.$$

If $q = 1$ we have only one case. If $q = 0$, we have to solve $10d + 5n + p = 25$. The number d can be 0, 1 or 2. Let us count the number of solutions in each case:

$d = 0$ gives 6 solutions, namely $n = 0, 1, 2, 3, 4, 5$.
$d = 1$ gives 4 solutions, namely $n = 0, 1, 2, 3$.
$d = 2$ gives 2 solutions, namely $n = 0, 1$.

Thus, the total number of ways to compose 25 cents is

$$1 + 2 + 4 + 6 = 13.$$

3.35. The probability that he got TOLEDO correct can be found as follows: the probability that T is correct is 1/20, since there are 10 letters and 2 positions for each letter. The probability that O is correct when T is correct is 4/9, since 9 letters remain, 4 are O's, and the position of the O does not matter. The probability that L is correct when T and O are correct is 1/16 (8 letters remain and L has 2 positions). Continuing this way we find:

$$\mathbf{Pr}\{\text{TOLEDO correct}\} = \frac{1}{20} \cdot \frac{4}{9} \cdot \frac{1}{16} \cdot \frac{1}{14} \cdot \frac{1}{12} \cdot \frac{3}{5}$$
$$= \frac{1}{201600}$$
$$\approx 4.96032 \cdot 10^{-6}$$

Proceeding in the same manner, we find that

$$\mathbf{Pr}\{\text{OHIO correct}\} = \frac{4}{10} \cdot \frac{1}{9} \cdot \frac{1}{8} \cdot \frac{3}{7}$$
$$= \frac{1}{420}$$
$$\approx 2.38095 \cdot 10^{-3}$$

The probability that he got both words correct is the probability that he got TOLEDO correct times the probability that he got OHIO correct given that he got TOLEDO correct. Proceeding as before we find

$$\mathbf{Pr}\{\text{both correct}\} = \frac{1}{201600} \cdot \frac{2}{4} \cdot \frac{1}{3} \cdot \frac{1}{2}$$
$$= \frac{1}{2419200}$$
$$\approx 4.1336 \cdot 10^{-7}$$

3.36. A path from the bottom left corner to the top right corner looks as follows:

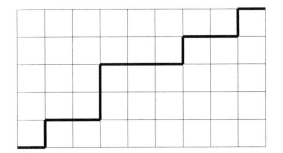

Note that the path is composed of $m+n$ segments of length equal to the length of the side of the squares of the grid. From these $m+n$ segments, n are vertical and m are horizontal (note that we have to go m rows up and n columns to the right). If we think of the path as a long line divided into $m+n$ segments, each different path in the grid is just a way to choose which of the segments are going to be horizontal and which are going to be vertical. Note that once we have decided which are horizontal, the rest are vertical. Thus, the number of paths equals the number of ways to choose m elements from a set of $m + n$ elements, i.e.

$$\binom{m + n}{m}.$$

Note that, since
$$\binom{m + n}{m} = \binom{m + n}{n},$$
another valid answer is

$$\binom{m + n}{n}.$$

3.37. First note that, for the two teams that played the first game, the number of losses equals the number of games—out of the 11— that they did not play. This is because they will sit out in a game if and only if they lost the previous one, and one of them won the last one against New York. For the New York team, this rule also holds, with the exception of the first game, in which New York sat out without having lost the previous game; but this is compensated by the fact that they lost the last game (i.e. we can think, as an abstraction, that they did not play the first game because they lost the last one). Therefore, for every team, the number of lost games equals the number of games they missed.

On the other hand, for each team we have

$$\#\{\text{games won}\} + \#\{\text{games lost}\} + \#\{\text{games missed}\}$$
$$= \#\{\text{total number of games}\}$$
$$= 11$$

Therefore we must have that, for each team

$$\#\{\text{games won}\} + 2 \cdot \#\{\text{games lost}\} = 11.$$

This means that each team won an odd number of games.

Let w_1, w_2, w_3 denote the number of games won by each team. Since, by hypothesis, they all won a different number of games, it must be that the w_i's are all different numbers. Thus the w_i's must be odd numbers, all different and must add up to 11 (since every game had a winner). By inspection, we see that the only possibility is 1,3,7. Also, by the last equation, we also know that the team that won 7 games lost 2, the team that won 3 games lost 4 and the team that won 1 game lost 5.

Note that this does not specify which team won 1,3 or 7 games. In fact, there are valid configurations in which the New York team won 1,3 or 7 games respectively, so this information cannot be inferred from the data given in the exercise.

3.39. The probability of getting a 12 when rolling two dice twenty-four times is

$$1 - \mathbf{Pr}\{\text{not getting a 12 in twenty-four rolls}\}$$
$$= 1 - \left(\frac{35}{36}\right)^{24}$$
$$= 1 - \left(\frac{5}{6}\right)^4 \cdot \left(\frac{7}{6}\right)^{24} \cdot \left(\frac{5}{6}\right)^{20}$$
$$= 1 - 0.508596$$

$$= \quad 0.491404$$
$$< \quad 0.5.$$

Chevalier de Méré was betting even money even though the game was slightly unfavorable for him. In fact, the payoff was only 98.2808 francs for every 100 francs he bet.

3.40. The probability of getting a 6 when rolling one die four times is

$$1 - \mathbf{Pr}\{\text{not getting a 6 in four rolls}\} = 1 - \left(\frac{5}{6}\right)^4.$$

The probability of getting a 12 when rolling two dice twenty-four times is, as calculated in the previous exercise,

$$1 - \left(\frac{35}{36}\right)^{24} = 1 - \left(\frac{5}{6}\right)^4 \cdot \left(\frac{7}{6}\right)^{24} \cdot \left(\frac{5}{6}\right)^{20}.$$

Since

$$\left(\frac{7}{6}\right)^{24} \cdot \left(\frac{5}{6}\right)^{20} \approx 1.05 > 1,$$

we have

$$1 - \left(\frac{5}{6}\right)^4 \cdot \left(\frac{7}{6}\right)^{24} \cdot \left(\frac{5}{6}\right)^{20} < 1 - \left(\frac{5}{6}\right)^4$$

3.41. For each toss, the number of possible outcomes is $2^5 = 32$. The number of outcomes that settle the game can be counted as follows: If the game is settled, that means that one person got a tail when all the others had heads or vice versa. Therefore, each person wins in 2 out of the 32 possible outcomes. Thus, the number of outcomes that settle the game is $2 \cdot 5 = 10$. Thus, the probability of having the game settled in the first toss is $5/16$. The probability of having it settled in the second toss is the probability that nobody won in the first toss times the probability that someone won in the second toss. This is, we have

$$\mathbf{Pr}\{\text{game settled in second toss}\} = \left(1 - \frac{5}{16}\right) \cdot \frac{5}{16} = \frac{55}{256}.$$

3.42. Here, best strategy means: If we were to repeat the experiment a large number of times, what would be the strategy that would minimize the average number of questions?

So we have to minimize the average number of questions. For a given strategy, the average number of questions is

$$1 \cdot \mathbf{Pr}\{\text{guessing in 1 question}\} + 2 \cdot \mathbf{Pr}\{\text{guessing in 2 questions}\}$$
$$+ \cdots + 7 \cdot \mathbf{Pr}\{\text{guessing in 7 questions}\}.$$

(Note that after 7 questions the solution is always known.)

Let us denote the color of the balls by their initials except for black, which will be denoted by X to avoid confusion with blue.

Notice that there is a strategy that always takes 3 questions. It goes as follows:

Q1: is it P, O, Y or X?

Assume that the answer is 'yes'. Then ask

Q2: is it Y or X?

Assume that the answer is 'no'. Then ask

Q3: is it P?

If the answer is 'yes', then the ball is purple; if the answer is 'no', then the ball is orange. If the answers to questions 1 and 2 are different to the ones above, the strategy is similar: always divide the set in which the chosen ball lies into two subsets and then ask if the ball is in one of the subsets.

Since we can do it in 3, we are not interested in any strategy that gives an average number of questions greater than 3.

Another strategy is asking all the colors in the list one by one. It makes sense that, if we want to minimize the average, we want to start with the most probable ones. It would not make sense to start with 'is it white?', because there is only probability 1/36 that this is the case (note that there are a total of 36 balls and only one is white).

For example, let us find the average in the following strategy:

Q1: is it P?
Q2: is it O?
Q3: is it Y?
Q4: is it X?
Q5: is it B?
Q6: is it G?
Q7: is it R?

(Of course, if the answer of any of these questions is affirmative, we do not have to ask the next question.)

The probability of guessing the color of the ball with the first question is the probability that the ball is purple, or 8/36. The probability of guessing with the second question is the probability that the ball is orange, or 7/36. Continuing in this fashion, we find that the average is

$$1 \cdot \frac{8}{36} + 2 \cdot \frac{7}{36} + 3 \cdot \frac{6}{36} + 4 \cdot \frac{5}{36} + 5 \cdot \frac{4}{36} + 6 \cdot \frac{3}{36} + 7 \cdot \frac{2+1}{36} = \frac{119}{36}.$$

(Note that the last summand is the probability that the ball is white or red, since in either case we will guess the color of the ball in 7 questions.)

Since $119/36 > 3$, this strategy is worse than the 'conservative' strategy we showed first.

There are many possible strategies, and to find the best one, one has to analyze them one by one. This process is quite tedious, but some of these strategies can be quickly ruled out. For example, it is clear that we must ask the most probable colors first.

The best strategy we have found is the following:

Q1: is it P or O?

If the answer is 'yes', in the next question we will ask if the ball is P. Either if it is P or O, we will guess the with 2 questions. If the answer is 'no', then ask

Q2: is it Y or X?

If the answer is 'yes',we will proceed as before and we will guess the answer with 3 questions. If the answer is 'no', then ask

Q3: is it B?
Q4: is it G?
Q5: is it R?

(Of course, if the answer to Q3 or Q4 is affirmative, the subsequent questions are not necessary.)

The average number of questions for this particular strategy can be found as follows: The probability of guessing the ball with only 1 question is clearly 0. The probability of guessing the ball with 2 questions is the probability that the ball is either P or O, which is 15/36. The probability of guessing the ball with 3 questions is the probability that the ball is either Y or X (which is 11/36), plus the probability that the answer to Q3 is affirmative (i.e. that the ball is B), which is 4/36. Thus, the probability of guessing with 3 questions is 15/36. The probability of guessing with 4 questions is the probability that the ball is G, i.e. 3/36 and the probability of guessing with 5 questions is the probability that the ball is R or W, i.e. 3/36. Putting all this together we obtain an average of

$$2 \cdot \frac{15}{36} + 3 \cdot \frac{15}{36} + 4 \cdot \frac{3}{36} + 5 \cdot \frac{3}{36} = \frac{102}{36}.$$

Note that the last quantity is less than 3. We encourage the reader to try other strategies and calculate the average number of questions for each strategy.

3.43. Let us follow the scheme given in PROBLEM 3.3.3. Let

$$F(x) = a_0 + a_1 x^1 + a_2 x^2 + a_3 x^3 + a_4 x^4 \cdots.$$

Notice that

$$3xF(x) = 3a_0 x + 3a_1 x^2 + 3a_2 x^3 + 3a_3 x^4 + 3a_4 x^5 \cdots,$$

and

$$x^2 F(x) = a_0 x^2 + a_1 x^3 + a_2 x^4 + a_3 x^5 + a_4 x^6 \cdots.$$

Grouping like powers of x, we obtain

$$F(x) - 3xF(x) + x^2 F(x)$$
$$= a_0 + (a_1 - 3a_0)x + (a_2 - 3a_1 + a_0)x^2$$
$$+ (a_3 - 3a_2 + a_1)x^3 + (a_4 - 3a_3 + a_2)x^4 + \cdots.$$

Now, since $a_j - a_{j-1} + a_{j-2} = 0$ for all $j \geq 2$, the expression above simplifies to

$$F(x) - 3xF(x) + x^2 F(x) = a_0 + (a_1 - 3a_0)x.$$

Since $a_0 = 2$ and $a_1 = 1$, we have

$$F(x)(1 - 3x + x^2) = 2 - 5x,$$

or

$$F(x) = \frac{2 - 5x}{1 - 3x + x^2}.$$

Manipulating the last expression, we find

$$F(x) = \left(1 - \frac{2}{\sqrt{5}}\right)\left[\frac{1}{1 - \frac{2}{3-\sqrt{5}}x}\right] + \left(1 + \frac{2}{\sqrt{5}}\right)\left[\frac{1}{1 - \frac{2}{3+\sqrt{5}}x}\right].$$

This can be written as

$$F(x) = \left(1 - \frac{2}{\sqrt{5}}\right)\sum_{j=0}^{\infty}\left(\frac{2}{3-\sqrt{5}}x\right)^j + \left(1 + \frac{2}{\sqrt{5}}\right)\sum_{j=0}^{\infty}\left(\frac{2}{3+\sqrt{5}}x\right)^j.$$

Therefore, the coefficient of x^j in the last expression is

$$\left(1 - \frac{2}{\sqrt{5}}\right)\left(\frac{2}{3-\sqrt{5}}\right)^j + \left(1 + \frac{2}{\sqrt{5}}\right)\left(\frac{2}{3+\sqrt{5}}\right)^j.$$

Since, on the other hand,

$$F(x) = \sum_{j=1}^{\infty} a_j x^j,$$

we must have

$$a_j = \left(1 - \frac{2}{\sqrt{5}}\right)\left(\frac{2}{3-\sqrt{5}}\right)^j + \left(1 + \frac{2}{\sqrt{5}}\right)\left(\frac{2}{3+\sqrt{5}}\right)^j.$$

3.44. Proceed as in the previous exercise. The solution is

$$a_j = \frac{7}{3} + \frac{5(-2)^j}{3}.$$

3.45. Proceed as in the two previous exercises. The solution is

$$a_j = 1 - 2^j.$$

3.46. The problem can be rephrased as follows: what is the probability that, if we choose 5 cards from a deck of 52, we will have exactly one face card? (the fact that we divide the deck into piles is irrelevant). The number of hands of 5 cards with only one face is

$$\#\{\text{ways to choose 4 non-face cards from 40}\}$$
$$\times \quad \#\{\text{ways to choose 1 face card from 12}\}.$$

(Note that there are 12 face cards and 40 non-face cards.)

Thus, the answer is

$$\frac{\binom{12}{1}\binom{40}{4}}{\binom{52}{5}} = \frac{703}{1666} = 0.421969.$$

3.47. Proceeding as in the previous exercise, the desired probability is

$$\frac{\binom{12}{1}\binom{40}{k-1}}{\binom{52}{k}}.$$

Note that, when $k > 40$, this probability is 1 (by the pigeonhole principle, for example).

3.48. The probability of choosing a white ball does depend on the distribution of the balls in the hats. The reason is that we first choose a hat, and then we choose a ball. For example, if at the beginning all the white balls are in one hat, the probability of choosing a white ball is 1/3, since we have 1/3 probability of choosing the 'right' hat. But if there are the same number of black and white balls in each hat, the probability of choosing a white ball is 1/2, since in this case, no matter which hat we pick at first, at the end we will choose a ball from a hat with 50% black balls and 50% white balls.

3.49. We will assume that each cow has probability 1/500 of having the disease, *independent of the fact that other cows may or may not have the disease.* Thus, the probability that some cow has the disease in a randomly selected group of 100 is 1/5.

For 100 cows, the expected number is

$$1 \cdot \mathbf{Pr}\{\text{all clean in } 100\} + 101 \cdot \mathbf{Pr}\{\text{some infected in } 100\}$$
$$= 1 \cdot \frac{4}{5} + 101 \cdot \frac{1}{5}$$
$$= \frac{105}{5}$$
$$= 21.$$

Thus, the expected number of tests for a total population of 5,000 cattle is $50 \cdot 21 = 1,050$.

3.50. At the beginning we have 1 cup of red dye dissolved in 81 cups of liquid. Thus, in each cup of the solution there are 1/81 cups of red dye. This implies that, after we remove one cup of the mixture and replace it with one cup of pure water, the amount of red dye that remains is

$$C_1 = 1 - \frac{1}{81} \quad \text{cups.}$$

Now we have $(1 - 1/81)/81$ cups of red dye per cup of solution. If we remove another cup of the mixture and replace it with one cup of pure water, the amount of red dye that remains is

$$C_2 = 1 - \frac{1}{81} - \frac{1 - 1/81}{81} = \left(1 - \frac{1}{81}\right)^2.$$

In general, if we have C_k cups of red dye in the mixture after doing this process k times, we will have $C_k/81$ cups of red dye per cup of mixture. Thus, if we remove another cup of mixture and replace it with water, the amount of dye that remains is

$$C_{k+1} = C_k - \frac{C_k}{81} = C_k \left(1 - \frac{1}{81}\right).$$

Thus, every time we do this process, the amount of dye left equals $(1 - 1/81)$ times the previous amount of dye. This implies that the amount of red dye left after doing this process k times is

$$C_k = \left(1 - \frac{1}{81}\right)^k \text{ cups.}$$

Since $(1 - 1/81) < 1$, every time we multiply by this number we obtain a smaller number. Thus, the amount (and, of course, the concentration) of dye is getting lower and lower. But, on the other hand, a nonzero number raised to a positive (finite) power is never zero, so the amount (or the concentration) is never zero after finitely many steps. As k becomes larger, C_k tends to zero (but is never zero for k finite, as discussed above). Thus there is no lower bound on the amount (or the concentration) of dye in the mixture. If we repeat this process enough times, we will have as low an amount of dye in the mixture as we desire.

3.52. Since we assume that these attributes are randomly (uniformly) distributed among people and are independent of each other, the proportion of the population that will be in the middle third in all three attributes is $(1/3)(1/3)(1/3) = 1/27$, or $100/27 \approx 3.7037\%$.

3.53. We will find the probability that the other has a pair or better (i.e. a pair, two-pairs, three of a kind, full house or poker) both when you have a pair and when all your cards are different.

Suppose that all your cards are different. Note that it does not matter which cards you have, as soon as they are all different, so we can assume that your cards are 1,2,3,4,5 of hearts (here, 1 stands for an ace). We want to find the probability that the other has a pair or better. Note that

$$\mathbf{Pr}\{\text{pair or better}\} = 1 - \mathbf{Pr}\{\text{all 5 cards different}\},$$

so let us instead find the probability that all her cards are different. The number of possible hands that she can have is $\binom{47}{5}$. To find how many of these hands contain no pair, we have to consider several cases:

If all her cards are in the set $\{1, 2, 3, 4, 5\}$, then there are 3^5 different hands with no pair (since there are 3 choices for 1, 3 for 2, etc).

If exactly four of her cards are in the set $\{1, 2, 3, 4, 5\}$, there are $\binom{5}{4}3^4\binom{8}{1}4^1$ different hands with no pair. This is because there are $\binom{5}{4}$ ways of choosing 4 numbers among $\{1, 2, 3, 4, 5\}$ and 3 suits per number, and $\binom{7}{1}$ ways of choosing 1 number among $\{6, 7, 8, 9, 10, J, Q, K\}$ and 4 suits per number.

If exactly three of her cards are in the set $\{1, 2, 3, 4, 5\}$, there are $\binom{5}{3}3^3\binom{8}{2}4^2$ different hands with no pair. The explanation is the same as above.

If exactly two of her cards are in the set $\{1, 2, 3, 4, 5\}$, there are $\binom{5}{2}3^2\binom{8}{3}4^3$ different hands with no pair.

If exactly one of her cards is in the set $\{1, 2, 3, 4, 5\}$, there are $\binom{5}{1}3^1\binom{8}{4}4^4$ different hands with no pair.

Finally, if none of her cards are in the set $\{1, 2, 3, 4, 5\}$, there are $\binom{8}{5}4^5$ different hands with no pair.

Thus, the probability that she has a pair or better when all your cards are different is

$$1 - \frac{1}{\binom{47}{5}}\left(3^5 + \binom{5}{4}3^4\binom{8}{1}4^1 + \binom{5}{3}3^3\binom{8}{2}4^2 + \binom{5}{2}3^2\binom{8}{3}4^3\right.$$
$$\left. + \binom{5}{1}3^1\binom{8}{4}4^4 + \binom{8}{5}4^5\right).$$

Now let us find the probability that all her cards are different when you have a pair. As before, we can assume that your cards are, for example, 1 of spades and 1,2,3 and 4 of hearts. There are again several cases. In each case, the number of hands is found using the same ideas as before.

If she has a 1 and exactly 3 cards in the set $\{2, 3, 4\}$, there are $2 \cdot 3^3 \cdot \binom{14}{1}4^1$ hands with no pair (we have 2 choices of 1,

3 choices for each card in $\{2, 3, 4\}$, $\binom{14}{1}4^1$ choices of number in $\{5, 6, 7, 8, 9, 10, J, Q, K\}$ and 4 choices of suit in this set.

If she has a 1 and exactly 2 cards in the set $\{2, 3, 4\}$, there are $2 \cdot \binom{3}{2}3^2 \cdot \binom{14}{2}4^2$ hands with no pair. The explanation of this is as above.

If she has a 1 and exactly 1 card in the set $\{2, 3, 4\}$, there are $2 \cdot \binom{3}{1}3^1 \cdot \binom{14}{3}4^3$ hands with no pair.

If she has a 1 and no cards in the set $\{2, 3, 4\}$, there are $2 \cdot \binom{14}{4}4^4$ hands with no pair.

Now, if she has no aces, we proceed similarly:

No aces and exactly 3 cards in the set $\{2, 3, 4\}$: $\binom{3}{3}3^3 \cdot \binom{14}{2}4^2$ hands.

No aces and exactly 2 cards in the set $\{2, 3, 4\}$: $\binom{3}{2}3^2 \cdot \binom{14}{3}4^3$ hands.

No aces and exactly 1 card in the set $\{2, 3, 4\}$: $\binom{3}{1}3^1 \cdot \binom{14}{4}4^4$ hands.

No aces and no cards in the set $\{2, 3, 4\}$: $\binom{14}{5}4^5$ hands.

This gives that the probability that she has a pair or better when you have a pair is

$$1 - \frac{1}{\binom{47}{52}}\left(2 \cdot 3^3 \cdot \binom{9}{1}4^1 + 2 \cdot \binom{3}{2}3^2 \cdot \binom{9}{2}4^2 + 2 \cdot \binom{3}{1}3^1 \cdot \binom{9}{3}4^3\right.$$

$$+2 \cdot \binom{9}{4}4^4 + \binom{3}{3}3^3 \cdot \binom{9}{2}4^2 + \binom{3}{2}3^2 \cdot \binom{9}{3}4^3$$

$$\left.+\binom{3}{1}3^1 \cdot \binom{9}{4}4^4 + \binom{9}{5}4^5\right).$$

Calculating these quantities, we find that the probability that the other has a pair or better when you have no pair is approximately 0.489636, and the probability that the other has a pair or better when you have a pair is approximately 0.495182. Thus, the second one is slightly higher.

Chapter 4

Problems of Logic

4.1. a) We start with E. The addition of E and O has yielded O. Hence, either $E = 0$, and nothing has been carried from the previous column, or $E = 9$, and there has to be a carry from the addition of N and R. Let us assume that the first case, $E = 0$, is true. This forces A to be 5 (How else can we get zero from $A + A$ in the fourth column?). Now we have

$$
\begin{array}{c}
1 \\
\text{D O N 5 L D} \\
+\ \text{G 0 R 5 L D} \\
\hline
\text{R O B 0 R T}
\end{array}
$$

Since we cannot have carrying from $N + R$, and $5 + 5 = 10$ forces a carry it follows that $N + R$ must be at most 8. In particular, since zero has been used and hence N has to be at least 1, we have that R is 7 or less. On the other hand, $D + G = R$, so R has to be at least 3. Now R cannot be 3, because then one of the pair D or G has to be 1 (and the other 2), and at the same time since $L + L$ produces 3 in the fifth column, L has to be 1 too.

If $R = 4$, then L has to be 2, and since $D + G = R = 4$, then one of the pair, D or G has to be 1 and the other 3. So by now all numbers from 0 to 5 are taken. This makes N at least 6 which means $N + R$ is at least 10, and we will have

79

to have a carry from this column. But that contradicts the $0 + O = O$ in the next column. So R cannot be 4.

Since 5 is already taken, this leaves us with 6 and 7 for R. Let us try $R = 6$. This forces L to be 3, and also means that $D + D$ in the first column cannot produce a carry, so D is 4 or less. At the same time, N cannot be more than 2, otherwise $N + R + 1$ forces an unwanted carry. It follows that D is not 1, because then $T = 2$, and the next choice for N would be too large. We cannot allow $D = 2$ either, because then G and T both have to be 4. So if $R = 6$, D has to be 4. Let us see what we have

$$
\begin{array}{c}
1\\
\text{D O N 5 3 4}\\
+\ \text{G 0 R 5 3 4}\\
\hline
\text{R O B 0 6 T}
\end{array}
$$

Next, we have $T = 8$ and $G = 2$. Now N has to be 1; but this cannot be, because then B would have to be 8, and 8 has already been taken.

The last choice for R is 7. Assuming this, N can only be 1. Furthermore, $L = 3$ and there has to be a carry from $D + D$, so that $L + L$ yields 7. Our display reads

$$
\begin{array}{c}
1\quad\ 1\\
\text{D O 1 5 3 D}\\
+\ \text{G 0 7 5 3 D}\\
\hline
\text{7 O B 0 7 T}
\end{array}
$$

Now D is greater than 5, since otherwise we will not have a carry from $D + D$. At the same time, since $D + G = R = 7$, D is less than 7. That leaves us with only 6 for D. But if $D = 6$ then G must be 1, and 1 is already taken.

We have run out of options for R, so we cannot proceed. The only thing we can do at the moment is to go back and change $E = 0$ to $E = 9$. This makes $A = 4$, and forces a carry from $L + L$. We have

$$
\begin{array}{r}
1\ 1\quad 1 \\
D\ O\ N\ 4\ L\ D \\
+\ G\ 9\ R\ 4\ L\ D \\
\hline
R\ O\ B\ 9\ R\ T
\end{array}
$$

Since $1 + D + G = R$, and none of D or G could be zero, R is at least 4. Now R cannot be 4, because $A = 4$. So R must be 5 or greater. Let us assume that $R = 5$. Then L would have to be 7, and we would have to have a carry from $D + D$. This makes D greater than 5 which cannot be, because $1 + D + G = R = 5$. Likewise, if $R = 7$ (or any other odd number), then D has to be greater than 5. But this time $1 + D + G = R$, forcing D to be 5. Then G has to be 1, and L has to be 8. And we have

$$
\begin{array}{r}
1\ 1\quad 1 \\
5\ O\ N\ 4\ 8\ 5 \\
+\ 1\ 9\ 7\ 4\ 8\ 5 \\
\hline
7\ O\ B\ 9\ 7\ T
\end{array}
$$

Clearly T must be zero. The only numbers left are 2, 3 and 6. With a little experimenting, we see that $N = 6$, $B = 3$ and $O = 2$. So the puzzle is solved as

$$
\begin{array}{r}
5\ 2\ 6\ 4\ 8\ 5 \\
+\ 1\ 9\ 7\ 4\ 8\ 5 \\
\hline
7\ 2\ 3\ 9\ 7\ 0
\end{array}
$$

We leave it to the reader to check that $R = 6$ or 8 do not yield any additional solutions.

c) The T in $TWELVE$ arises from carrying and since

$$SEVEN + EIGHT < 1000000 + 1000000 = 2000000,$$

T must be 1. Furthermore, notice that in the third column from the left, $E + I$ yields E. Hence I is either zero or 9. Let us first assume that $I = 0$.

In the last column (counting from the left), $N + T = N + 1$ yields E. Since E cannot be zero, we cannot have a carry from the right, and we must have $N + 1 = E$. We have

$$
\begin{array}{r}
\text{S\ \ E\ V\ E\ N} \\
+\ \text{E\ \ 0\ G\ H\ 1} \\
\hline
\text{1\ \ W\ E\ L\ V\ E}
\end{array}
$$

We see that E has occurred five times in the above addition. Hence, knowing E would give a lot of information. So it is worthwhile to try different values for E.

Since $N + 1 = E$, and N is at least 2, we start with $E = 3$. Then

$$
\begin{array}{r}
\text{S\ \ 3\ V\ 3\ 2} \\
+\ \text{3\ \ \ 0\ G\ H\ 1} \\
\hline
\text{1\ \ W\ 3\ L\ V\ 3}
\end{array}
$$

Now S has to be 7, 8 or 9, so that $S + 3$ in the second column forces a carry. This implies that $W = 0, 1$ or 2. But all of these numbers are already taken. This means that we have to try another number for E. If $E = 4$, we have

$$
\begin{array}{r}
\text{S\ \ 4\ V\ 4\ 3} \\
+\ \text{4\ \ \ 0\ G\ H\ 1} \\
\hline
\text{1\ \ W\ 4\ L\ V\ 4}
\end{array}
$$

Obviously S has to be at least 6 to force a carrying operation. On the other hand, S cannot be 6, 7 or 9 (otherwise $W = 0, 1$ or 3, and all these are already taken). This forces S to be 8 and W to be 2. Now, since all the numbers from 0 to 4 have been taken, $V + G$ has to be at least 11. But we cannot have a carry from this column. So, assuming $E = 4$ leads to a contradiction.

The next natural choice is $E = 5$. Substituting this we get

$$
\begin{array}{r}
\text{S\ \ 5\ V\ 5\ 4} \\
+\ \text{5\ \ \ 0\ G\ H\ 1} \\
\hline
\text{1\ \ W\ 5\ L\ V\ 5}
\end{array}
$$

The possible values for S are 7 and 8. If $S = 7$, then $W = 2$. Since $V + G$ cannot produce a carry, one of V or G has to be less than 5. The same is true for V and H (why?). Since there is only one such number left, namely 3, $V = 3$. Then $H = 8$, and G has to be 6. But we are in trouble here, because $3 + 6$ with a carry yields 0 and a carry. We have a contradiction so we proceed to the next choice for S.

If $S = 8$, we get $W = 3$. Similar reasoning as in the previous paragraph implies that $V = 2$, $H = 7$ and $G = 6$. Then $L = 9$, and the puzzle is solved as

$$
\begin{array}{r}
8\,5\,2\,5\,4 \\
+\,5\,0\,6\,7\,1 \\
\hline
1\ \ 3\,5\,9\,2\,5
\end{array}
$$

Going back to the beginning of the solution, we see that we have not explored the possibility that $I = 9$. As a matter of fact, this gives us another solution as follows:

$$
\begin{array}{r}
6\,3\,7\,3\,2 \\
+3\,9\,8\,4\,1 \\
\hline
1\,0\,3\,5\,7\,3
\end{array}
$$

We leave the details to the reader.

e) This puzzle has many solutions. Among them are

$$
\begin{array}{r}
3\,4\,2 \\
+\,1\,3\,5\,0 \\
\hline
1\,6\,9\,2
\end{array}
$$

and

$$
\begin{array}{r}
4\,6\,2 \\
+\,8\,4\,5\,0 \\
\hline
8\,9\,1\,2
\end{array}
$$

g) There are many solutions, such as

$$
\begin{array}{r}
1\ 7\ 3 \\
+\ 2\ 9\ 5 \\
\hline
4\ 6\ 8
\end{array}
$$

and

$$
\begin{array}{r}
2\ 8\ 9 \\
+\ 4\ 6\ 1 \\
\hline
7\ 5\ 0
\end{array}
$$

i) By trial and error one gets these as possible solutions to the puzzle

$$
\begin{array}{rcl}
63 \times 154 &=& 9702 \\
54 \times 168 &=& 9072 \\
59 \times 136 &=& 8024 \\
26 \times 345 &=& 8970
\end{array}
$$

4.2.   a) First let us label $*$'s with letters so that we can refer to them.

$$
\begin{array}{r}
A\ B\ C \\
D\ E \\
\hline
F\ G\ H\ I\ 1
\end{array}
$$

All different letters stand for different digits. We have to find the operation first. The addition of a two digit integer to a three digit integer does not produce a five digit integer, much less if we subtract or divide these two numbers. Thus the operation must be multiplication.

Now $C \times E$ must produce a 1 in the lowest position of the product. We have $1 \times 1 = 1$, $9 \times 9 = 81$, and $3 \times 7 = 21$. Since C and E have to be different, either $C = 3$ and $E = 7$, or $C = 7$ and $E = 3$.

Let us see how large the result of the product can be. If $A = 9$ then D has to be 8 or less. In this case

$$ABC \times DE < 999 \times 90 < 90,000.$$

If $D = 9$ then A is 8 or less, and $ABC \times DE < 900 \times 99 <$ $90,000$. In any case F is 8 or less.

If $F = 8$, since we cannot use 8 twice, then either $D < 8$ or $A < 8$. In either case, $ABC \times DE < 80,000$. Thus F cannot be 8. Since 7 is already taken, we will try $F = 6$ next. With this assumption, we try to find all possible choices for the pair A and D.

If we assume $A = 9$, then 8 is too large for D ($900 \times 80 =$ 72000), 6 and 7 are taken, and 5 is too small ($999 \times 59 <$ $60,000$). Hence A is not 9. Similarly A cannot be 8 either. The cases $D = 8$ or $D = 9$ are dismissed the same way. But not both A and D can be less than 7, because $699 \times 69 <$ $49,000$ (remember that 7 is already taken). We have run out of options for the pair A and D, so F cannot be 6.

Next we check $F = 5$. We want to find all the possible combinations for the pair A and D. Let us start with $A = 9$. Then again 8 is too large; 7 is already taken, but $D = 6$ is a possibility. The next choice, 4, is already too small. Likewise if $A = 8$ the only choice for D would be 6. For $A = 6$, we get $D = 9$ or 8. Also $A = 4$ is too small because $499 \times 99 < 50,000$. These results have been summarized in the following table:

$$A = 9 \quad \text{and} \quad D = 6.$$
$$A = 8 \quad \text{and} \quad D = 6.$$
$$A = 6 \quad \text{and} \quad D = 9.$$
$$A = 6 \quad \text{and} \quad D = 8.$$

On the other hand, $\{C, E\} = \{3, 7\}$. Combining this with the above list we get

$$
\begin{array}{llllll}
A= 9; & C= 3; & D= 6; & E= 7 & \text{or} \\
A= 9; & C= 7; & D= 6; & E= 3 & \text{or} \\
A= 8; & C= 3; & D= 6; & E= 7 & \text{or} \\
A= 8; & C= 7; & D= 6; & E= 3 & \text{or} \\
A= 6; & C= 3; & D= 9; & E= 7 & \text{or} \\
A= 6; & C= 7; & D= 9; & E= 3 & \text{or} \\
A= 6; & C= 3; & D= 8; & E= 7 & \text{or} \\
A= 6; & C= 7; & D= 8; & E= 3.
\end{array}
$$

If we start with the first row, we have $9B3 \times 67 = 5GHI1$. Now the choices for B are $0, 1, 2, 3, 4$ or 8. So it is a matter of checking all these possibilities. Overall, we have 48 multiplications. For example, $917 \times 63 = 57771$ is not acceptable as a solution, because 7 has been repeated four times. After checking all these multiplications, we get $927 \times 63 = 58401$ as a solution.

If we want to look for more solutions, we have to also examine the cases $F = 4, 2$ and 1. For each choice, we find a table similar to the above table, and then we do the multiplications to find acceptable solutions. We leave it as an exercise for the reader to do this. The additional answers are

$$917 \times 53 = 48601 \text{ and}$$
$$823 \times 57 = 46911.$$

c) As in part a) the operation must be multiplication. By trial and error, one gets the following as a solution

$$
\begin{array}{r}
3\,6\,7 \\
\times \quad 5\,2 \\
\hline
1\,9\,0\,8\,4.
\end{array}
$$

e) Since the multiplication of two three-digit numbers produces at least a five digit number, the operation here must be addition. This problem has many solutions. Here is one.

$$
\begin{array}{r}
6\,5\,2 \\
+\,4\,3\,7 \\
\hline
1\,0\,8\,9
\end{array}
$$

4.3. In the Gregorian Calendar, there are 52 weeks and 1 day in a normal year. Leap years have 1 extra day, February 29th. Hence, if year x starts on a Saturday, year $x + 1$ will start on a Sunday if x is a normal year; otherwise it will start on Monday.

Recall that all years which are divisible by 4 are leap years, except the years which are divisible by 100, in which case they are leap years only if they are divisible by 400. So year 2100 is not a leap year, whereas year 2000 is.

On the other hand, a period of 400 years has exactly 20871 weeks (refer to Exercise 12). Thus years x and $x + 400$ start with the same day of the week, no matter which year x is. This implies that to know the relative frequencies by which New Year's Day falls in different days of the week, it is enough to know these frequencies only in a period of 400 years.

It would be a terribly boring and tedious job if we were to sit down and write 400 consecutive years and their corresponding New Year's Days. Fortunately, there are steps that we can take to make the job easier.

First it is obvious that each period of 400 years has one and only one year which is divisible by 400. If it were not for this year, we could divide the interval of 400 years into 4 subintervals of 100 years, and then find how many times each particular day becomes New Year's Day in the first 100 year. Then since all the other 3 subintervals would have had the same pattern, except that each starts on a different day, by renaming the days of the week we could find the number of times each day becomes New Year's Day in the other subintervals. Finally by adding the figures we could answer the question.

We can do this if we choose the period so that the "troubling" year is at the end of one of the subintervals, preferably at the end of the interval itself. For example, the interval from 2001 to 2400 will do.

Before starting to count, let us refer to each day of the week by numbers $1, 2, \ldots, 7$, assuming that the year 2001 starts with day 1. We will give the appropriate names at the end.

To make counting even easier, we notice that in each interval of 28 years which does not include the years which are divisible by 100, New Year's Day fall on each day of the week exactly 4 times. We leave it to the reader to either prove this or convince himself/herself by counting. Thus by the year 2084 each day will have been New year's day exactly for $12 = 3 \times 4$ times. We only have to include the last sixteen years which we do simply by counting. Knowing that the years 2001 and 2085 start on the same day we have

$$1, 2, 3, 4, 6, 7, 1, 2, 4, 5, 6, 7, 2, 3, 4, 5$$

as New Year's Days for the last 16 years. Notice that the next period of one-hundred years will start with day 6.

We summarize all the result in the following table.

n	1	2	3	4	5	6	7
$A_1(n)$	14	15	14	15	14	14	14

Here $A_i(n)$ denotes the number of times New Year's Day falls on day n during the ith subinterval.

As we mentioned before, the next hundred years starts with day 6. Hence, by relabelling the above table, we get

n	6	7	1	2	3	4	5
$A_2(n)$	14	15	14	15	14	14	14

Similarly we get the following tables.

n	4	5	6	7	1	2	3
$A_3(n)$	14	15	14	15	14	14	14

n	2	3	4	5	6	7	1
$A_4(n)$	14	15	14	15	14	14	14

Adding the corresponding figures for each day we get Table 4.1, in which $A(n)$ refers to the number of times day n will be New Year's Day in the period 2001 to 2400.

n	1	2	3	4	5	6	7
$A(n)$	56	58	57	57	57	56	58

Table 4.1

Now it only remains to give appropriate name to each of the numbers. This can be done easily by checking a calendar to see that the year 2001 starts on a Monday, so in the table Table 4.1 Saturday is day 6, and Sunday is day 7. Thus, New Year's Day falls more often on Sunday than on Saturday.

4.4. After having done the previous exercise, this should be easy. Each year has 11 days which are the 30th of a month. For each of these we can use the Table 4.1 by giving the appropriate names to the numbers 1 to 7. For example, the calendar tells us that January 30th of the year 2001 is on Tuesday. Thus Tuesday is day 1, and we know for example that in the course of 400 years, Tuesday is the 30th of January 56 times, and so on. Doing this for all the other days and adding the result we can answer the question. There is only one problem that we will explain shortly.

Recall that, in each leap year, the extra day is February 29th. Hence, in each leap year, for the days after February 29th the leap has already occurred, whereas for the days before February 29th the leap is yet to happen. So what we called the "troubling" year will be different in the two cases. For the latter it is the same year as before (i.e. previous problem), and for the former it would be the years such as 1999 or 2399. For example, since March 30th in the year 2000 is a Thursday, day 1 in the Table 4.1 will be Thursday, and so on.

Mon.	Tue.	Wed.	Th.	Fri.	Sat.	Sun.
630	625	629	628	626	628	624

Thus, Monday is the day on which the 30th of a month falls most.

4.5. In her first move, the first player should place the center of the poker chip at the center of the table. After that, each time the second player makes a move, the first player should place her chip in exactly the symmetric position with respect to the center of the table. This way each time that the first player completes her move, the position of the poker chips on the table are symmetric with respect to the center. Hence, if the second player has a move,

the first player will also have a move. So she has to be the player
who puts the last chip.

4.6. The first and third statements imply that the conductor's last
name is not Smith. Also from the sixth statement it follows that
the waiter's last name is not Pistilgaglioni. By the second statement we know that the conductor lives halfway between New York
and Washington, so neither of the passengers who live in these
two locations is the passenger who lives nearest to the conductor.
On the other hand Brown, who earns $2000 per month, cannot
earn three times the conductor because 2000 divided by 3 is less
than 700, which is less than the minimum wage. So Brown does
not live nearest to the conductor, and since he does not live in
New York either, he must live in Washington. So he has the same
last name as the conductor.

By now we know that the conductor's last name is Brown, and
the waiter's last name is not Pistilgaglioni, so his name must
be Smith. Then the engineer cannot have any other name than
Pistilgaglioni.

4.7. Theodore's first and last statements are either both false or both
true. Since each student has said exactly one false statement,
both of these must be true. So Theodore is not the culprit.

Now we examine David's statements. In his third statement, he
claims that Theodore has stolen the purse. We know that this is
false. So we know that his other two statements are true, including the statement claiming that he, David, was not acquainted
with Margaret before the beginning of this school year. From
this we conclude that Margaret is lying when she says that David
has known her for many years. So she is telling the truth in
her other two statements, including when she says that Judy has
taken the purse.

4.8. a) Since $(ATOM)^2 = ****ATOM$, M can only be 1, 5, or 6.
Let $M = 1$. Then $O = 0, 2, 3, \ldots$ or 9. Examining all these
choices we see that $O = 0$ is the only acceptable one. For
example, if $O = 2$, then $(AT21)^2 = ******41$, which is

clearly not acceptable. Now $T = 2, 3, \ldots$ or 9. But we see that T cannot be 2, because $(A201)^2 = *****401$. Similarly T cannot be any of the values $3, 4, \ldots$ or 9. Thus, we have to change the assumption $M = 1$.

Now let us assume $M = 5$. Since $(10a+5)^2 = 100(a^2+a)+25$, we have that $(ATO5)^2 = ******25$. It follows that $O = 2$. In exactly the same way, using algebra, we get $T = 6$ and $A = 0$. But we do not write 0 at the beginning of a number, so M cannot be 5.

The only remaining option for M is 6. Assuming this we examine all the possible values for O. We get that $O = 7$. Similarly, $T = 3$, and $A = 9$. So the puzzle has been solved as

$$
\begin{array}{r}
9376 \\
\times 9376 \\
\hline
56256 \\
65632 \\
28128 \\
84384 \\
\hline
87909376
\end{array}
$$

b) The answer is: $ABC = 246$.

c) The answer is:

$$
34 + 56 = 90
$$
$$
72 + 10 = 82
$$
$$
56 + 10 + 82 = 148
$$

4.9. It follows from the first statement that Nod and Schmotzky have different mothers, so they cannot be the same person. Naturally, Nod's first name is not Schmotzky. On the other hand, Blinken and Schmotzky are different people by the fourth statement. So Schmotzky's last name must be Winken. He also must be 12 years old, because he began first grade when he was 7, and now he is beginning sixth grade.

From the last statement, we cannot conclude that Blinken and Plotzky are different people. And the assumption that Plotzky's last name is Blinken is consistent with all the other statements. However, with this assumption we do not have enough information to find Blotzky Nod's age. So this problem is solvable only if Plotzky and Blinken are different people. Then the three boy scouts are as follows: Schmotzky Winken 12, Blotzky Nod 13 and Plotzky Blinken 13.

4.10. For simplicity let us refer to the narrator by F, to his son by S and to his father by G. Also assume that $A(G)$, $A(F)$ and $A(S)$ indicates the age of the corresponding person. F says, "When I am as old as my father is now, I will be five times as old as my son is now." In simple words this means that G is five times older than S.

$$A(G) = 5A(S).$$

Then F adds, "But at that time my son will be eight years older than I am now." In other words the age difference between F and S is 8 years less than the age difference between F and G. So

$$A(G) - A(F) = A(F) - A(S) + 8.$$

Simplifying the above and using the fact that $A(G) = 5A(S)$, we get

$$A(F) + 4 = 3A(S).$$

This means that $A(G) + A(F) + 4 = 5A(S) + 3A(S) = 8A(S)$. On the other hand, F in his last sentence notes that the sum of the ages of himself and his father is 100. So $A(G) + A(F) + 4 = 104 = 8A(S)$, and $A(S) = 13$. Knowing this, we can also find that $A(F) = 3(13) - 4 = 35$, and $A(G) = 100 - 35 = 65$.

4.11. The question is how many times we can switch the minute and the hour hands and still get a valid time. For example, if the clock shows 1 o'clock, after switching the hands, the hour hand will be exactly on 12, and the minute hand will be on 1, which is not a valid time because the hour hand would have to be a little past the top when the minute hand is past the top.

Now suppose h is the time. It also coincides with the position of the hour hand in the clock (we make the convention that 12:00

is the hour zero.) Also let L denote the position of the minute hand. Then

$$h = k + \frac{L}{12},$$

where $k = 0, 1, \ldots, 11$. If we switch the hands, and the clock still shows a valid time, then the following must be true too.

$$L = k' + \frac{h}{12},$$

where $k' = 0, 1, \ldots, 11$. Eliminating L between the above two equations we get

$$h = \frac{144k + 12k'}{143}.$$

This gives 12×12 combinations. We have to take into the account that the beginning of the cycle, corresponding to $k = k' = 0$, and the end of the cycle, corresponding to $k = k' = 11$, are actually both the same time, which is 12 o'clock. Hence as time passes the clock would show 143 valid times.

We suggest to the dubious reader to make a table of h values for all the 144 combinations of k and k', to see that the only repeated values are for $k = k' = 0$ and $k = k' = 11$. We believe that after computing the first two rows and the first two columns he or she will be convinced.

4.12. In the Gregorian Calendar, there are 52 weeks and one day in each year, except for leap years which have 52 weeks and two days. Every year which is divisible by 4 is a leap year except if it is divisible by 100 in which case it is a leap year only if its divisible by 400. For example, the year 2000 is a leap year whereas the year 1900 is not. So a period of 400 years consists of 400×52 weeks and 497 days which makes it exactly 20871 weeks. Hence the years 2000 and 2400 start on the same days of the week, and so do the years 2001 and 2401 and so on.

Let us pick the the year 2000. One can check with a calendar that it starts on a Saturday. In the next century, there are 100×52 weeks and 125 days. This is short of 5218 weeks by one day.

It follows that the year 2100 should start with a Friday. All of the following three centuries have one day less in them, because none of them starts with a leap year. So they are each short of 5218 complete weeks by two days. Hence the year 2200 starts on Wednesday, the year 2300 starts on Monday and 2400 as expected starts on Saturday. The same pattern repeats for the years after 2400 and before 2000. So a century can begin with any of Monday, Wednesday, Friday or Saturday.

Let us remark that, had we started the analysis with any other century, the result would have been the same. The year 2000 is only the most convenient.

4.13. It is possible to cover the whole board in 64 moves. To find a way to do so, we make an ordinary chess board of 64 cells. We move the knight on the board till there are no more moves. We label the remaining squares by $a, b, c, \ldots$. Then we try to add them to the route. We illustrate this through one example. Suppose that we have the following route.

42	21	54	9	40	19	52	7
55	10	41	20	53	8	39	18
22	43	24	63	30	59	6	51
11	56	31	60	27	62	17	38
32	23	44	25	58	29	50	5
45	12	57	28	61	26	37	16
a	33	2	47	14	35	4	49
1	46	13	34	3	48	15	36

We have covered everything except square a. Notice that a commands the square 57, and 63 commands the square 56. So if we change the sequence to $1, \ldots 56, 63, \ldots 57$, we can add a to the end of the route. After renumbering we have

42	21	54	9	40	19	52	7
55	10	41	20	53	8	39	18
22	43	24	57	30	61	6	51
11	56	31	60	27	58	17	38
32	23	44	25	62	29	50	5
45	12	63	28	59	26	37	16
64	33	2	47	14	35	4	49
1	46	13	34	3	48	15	36

Sometimes one might need to change the sequence a few times before one can add any new cell to the route. As an exercise, the reader can try to make the above path re-entrant (i.e. the knight finishes in the same square that it started.). For further reading on the subject see "Mathematical Recreations and Essays" by W. W. Rouse Ball.

4.14. Let P be the price of one turkey in cents. Then $72P = a679b$, where a and b stand for the erased digits. Here we are assuming that P is a whole integer. Let us also refer to $a6792b$ by X. Now X is divisible by 8 since 72 is. But since $X = a6(1000) + 79b$, and 1000 is divisible by 8, $79b$ should also be divisible by 8. It follows that $9b$ must divide 4, so $b = 2$ or 6. Between the two numbers 792 and 796, only 792 is divisible by 8. Thus $b = 2$.

On the other hand, X also divides 9. This implies that the sum of its digits should divide 9. So $a + 6 + 7 + 9 + 2 = a + 24$ must be divisible by 9. From here it follows that $a = 3$. Therefore P, the price of one turkey, is $36792/72 = 511$ cents or five dollars and eleven cents.

4.15. Simple counting shows that there are 49 ways of forming 50 cents using quarters, dimes, nickels, and pennies. Here we do the counting methodically.

Let $N_i(k)$, $i = 1, 5, 10, 25$, denote the number of ways that one can make k cents, using coins of value less than or equal to i. Here we want to find $N_{25}(50)$. Obviously, $N_1(k) = 1$, and $N_5(5k) = k + 1$.

In order to calculate $N_{25}(50)$, we break it into to three subproblems. First, we may not use any quarters, in which case there are $N_{10}(50)$ ways of forming 50 cents. Or we may use one quarter and make the other 25 cents using dimes, nickels and pennies. So the number in this case is $N_{10}(25)$. Finally one may use two quarters, and of course there is only one way of forming 50 cents using two quarters. Now everything is covered. Hence,

$$N_{25}(50) = N_{10}(50) + N_{10}(25) + 1 \qquad (4.1)$$

Now we have to calculate $N_{10}(50)$ and $N_{10}(25)$. Similarly, to make 50 cents using dimes, nickels, and pennies, we may use no dimes, one dime, two dimes, etc. So

$$
\begin{aligned}
N_{10}(50) &= N_5(50) + N_5(40) + \cdots + N_5 + 1 \\
&= 11 + 9 + 7 + 5 + 3 + 1 \\
&= 36
\end{aligned}
$$

and,

$$
\begin{aligned}
N_{10}(25) &= N_5(25) + N_5(15) + N_5(5) \\
&= 6 + 4 + 2 \\
&= 12.
\end{aligned}
$$

Substituting these into (4.1) we get

$$N_{25}(50) = 36 + 12 + 1 = 49.$$

In this way we can calculate $N_{25}(k)$ for any number of dollars. For example,

$$N_{25}(100) = N_{10}(100) + N_{10}(75) + N_{10}(50) + N_{10}(25) + 1$$

In the above, we already know the last three terms. So we only have to find the first two, which is done similarly. Next we can find $N_{25}(125)$ and etc.

It is possible to calculate a formula for $N_{25}(50k)$. In fact

$$N_{25}(50k) = \frac{100k^3 + 135k^2 + 53k + 6}{6}.$$

Since the process of calculating the above formula is rather tedious, we skip it. But it is worthwhile to mention that once we have the formula, we can actually prove it by induction. First we check the formula for $k = 1$ (since we have just found $N_{25}(50)$ it is easy to check that the formula is correct in this case). Then we prove that if it is valid for k, then it must be valid for $k + 1$ as well. In this manner we cover all the integers. [The relation $N_{25}(50(k + 1)) = N_{10}(50(k + 1)) + N_{10}(50k + 25) + N_{25}(50k)$ will be helpful in the induction step.]

4.16. The ladies consume a total of 14 sodas, so their husbands must have drunk the rest of the sum which is 30 sodas. Let x, y, z and w denote the number of sodas consumed by Mrs. Mergetroyd, Mrs. Ahmenhotep, Mrs. Ataturk and Mrs. Herkimer respectively. We have

$$x + 2y + 3z + 4w = 30.$$

There are 4 choices for x. Having chosen x, we will have 3 choices for y. Then having chosen x and y, there will only be two choices remaining for z. By knowing the values of x, y, z, the value of w will also be known. So we have a total number of $4 \times 3 \times 2 = 24$ possible solutions that we can check with the above equation. It is a good idea to start with some guess and try to correct the guess. The values which make the above sum, $x + y + z + w$, maximum or minimum (whichever is closer to 30) are good places to begin. After finding one solution we have to show that the solution is unique, because otherwise we cannot find the ladies' last names.

Here we try to perform both of these procedures at the same time by narrowing down the search. Notice that $x+y+z+w = 14$. If we subtract this from the previous equation we get $y + 2z + 3y = 16$. If $z = 4$ then $y + 3w = 8$. This forces $w = 2$, but then the least value y can take would be 3 in which case $y + 3w$ is still greater than 8. Now if 4 is too big for z, then z cannot be 5 either. In the same way 4 and 5 are too big for w too. So z and w can only take the values 2 and 3 which leaves 4 and 5 for x and y. Now we only have four options to check. Doing this we get $x = 5$, $y = 4$, $z = 3$ and $w = 2$. The four ladies present were Selma Herkimer, Hyacinth Ataturk, Lucinda Ahmenhotep and Myrtle Mergetroyd.

4.17. The following table is useful.

$$
\begin{array}{llll}
72 & = & 1 \cdot 1 \cdot 72 & \quad 1 + 1 + 72 = 74 \\
72 & = & 1 \cdot 2 \cdot 36 & \quad 1 + 2 + 36 = 39 \\
72 & = & 1 \cdot 3 \cdot 24 & \quad 1 + 3 + 24 = 28 \\
72 & = & 1 \cdot 4 \cdot 18 & \quad 1 + 4 + 18 = 23 \\
72 & = & 1 \cdot 6 \cdot 12 & \quad 1 + 6 + 12 = 19 \\
72 & = & 1 \cdot 8 \cdot 9 & \quad 1 + 8 + 9 = 18 \\
72 & = & 2 \cdot 2 \cdot 18 & \quad 2 + 2 + 18 = 22 \\
72 & = & 2 \cdot 3 \cdot 12 & \quad 2 + 3 + 12 = 17 \\
72 & = & 2 \cdot 4 \cdot 9 & \quad 2 + 4 + 9 = 15 \\
72 & = & 2 \cdot 6 \cdot 6 & \quad 2 + 6 + 6 = 14 \\
72 & = & 3 \cdot 3 \cdot 8 & \quad 3 + 3 + 8 = 14 \\
72 & = & 3 \cdot 4 \cdot 6 & \quad 3 + 4 + 6 = 13 \\
\end{array}
$$

We know that Sam knows the product of the boys' ages, which is 72. He also knows the sum of their ages, which we do not know. However, we know that the problem is indeterminate. In the above table, we have written 72 as product of three integers in all possible ways. We have also included the sum of the integers in each row. Now 14 is the only number which appears more than once. Had the street number been any of the other numbers, Sam would have had enough information to find the boys' ages. So the street number is 14, and then his friend has three boys at ages either 2, 6, 6 or 3, 3, 8. Then he says that he hopes that one day his oldest son will be quarterback in the U.S.C. football team. So he has one oldest son, and it follows that his boys have ages $3, 3$ and 8.

4.18. The game of tic-tac-toe is played on a board of nine squares (3 in each row). The two players take turns and mark one of the squares each time. The player who first marks 3 squares in a row (horizontal, vertical or diagonal) wins the game. We will attempt to illustrate through some examples that if both players play the game wisely, it will end in a draw.

Let us refer to the players by A and B, A being the first player. We also assume that A marks the center square in his first move, and B marks one of the corners. Now A basically has four different options as follows (he actually has seven but because of symmetry they reduce to four).

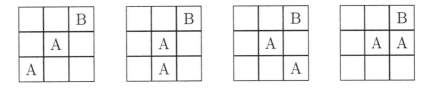

In the first case (from the left), B is forced to play in one of the corners, otherwise A can win in his fourth turn (why?). Notice that because of the symmetry it does not matter which one of the remaining corners B will choose. In the other three cases too, B has only one move which does not result in his defeat. And, except for the last case, the same is true for A in his second turn. And this continues till all these games end in a draw.

Since the last case seems a little different (A has more options in his second move), we follow it. After the second player has made his (forced) move we have the following

		B
B	A	A

Now the only constraint on A is that he should not mark the lower right corner square because then B will mark the upper left corner, and regardless of A's next move he will win the game. But any other move will only lead to a draw.

4.19. This problem is done similarly to the previous exercise.

4.24. Let us refer to the first player as A and to the second as B. We first devise a winning strategy for A.

We analyze the game backward. Player A wins if in her last turn the sum is between 90 and 99. Then she can add the required amount and make the sum 100. In order to have this situation, in her next to last turn, A should have made the sum 89 so that no matter what number B adds it will be between 90 and 99. But she can do this only if the previous sum is between 79 and 88. So in her second to the last turn she should have made the sum 78.

Continuing the same way, we see that in order for A to win the game, each time B should end up with the following sums in his turn. Numbers are written in reverse order.

$$89, 78, 67, 56, 45, 34, 23, 12, 1$$

So A writes down 1. Then B writes his number (between 1 and 10). A writes the difference so that the sum is 12. Again regardless of what B writes, next time A makes the sum 23 and so it goes.

If A, the first player, plays the game wisely, then B does not have any winning strategy. But the first time A makes a mistake, B can follow the above strategy and win the game. For example if in her first turn, A writes 2 instead of 1, then B should add 10 to it to make the sum 12. Now the situations have reversed and B wins the game.

4.25. "I completely disagree with you."

4.26. There is one obvious way of distributing the money so that it complies with the conditions of the problem, and that is to divide it equally among all ten people. We show that this is the only way.

Regardless of how we distribute the money, there is always somebody who gets the least share. Let us refer to him by P. There might be more than one such person, in which case we pick one and call him P. In fact we demonstrate that all of them must get the same amount as P. Since P has received the average share of his two neighbors, if one of them has a bigger share than P, then the other should have a smaller share. But P was supposed to have received the least share. So both neighbors must have the same share as P. Continuing the same way, we conclude that everybody gets an equal share.

Notice that the amount of money or the number of people are irrelevant for this problem. The key factor is that people are sitting at a round table so that each person has two neighbors.

4.27. First 32118 factors to $2 \cdot 3 \cdot 53 \cdot 101$. From the third statement we get that $C \geq 2$. Since $C < A < 100$, the only choice for A is 53. On the other hand we can have several legitimate choices for C and l, which we list below.

$$C = 2, \qquad l = 303;$$
$$C = 3, \qquad l = 202;$$
$$C = 6, \qquad l = 101.$$

4.28.   a)

$$
\begin{array}{r}
\mathrm{THE} \\ \hline
\mathrm{SHE\,|FEARS} \\
{*}\,{*}\,{*} \\ \hline
{*}\,{*}\,{*}\,{*} \\ \hline
\mathrm{TALK} \\
{*}\,{*}\,{*}\,{*} \\ \hline
{*}\,{*}\,{*}\,{*} \\ \hline
\end{array}
$$

From $THE \times SHE = FEARS$, it follows that S is the same as the digit in the lowest position in E^2. Hence S is $0, 1, 4, 6$ or 9. We dismiss $S = 0$ or 5, because then $E = S$.

Suppose $S = 9$. Then $E = 3$ or 7. Also notice that $T \times SHE$ gives the third line in the division, which is a three digit number. This leaves us with no other choice than 1 for T. Anything else would be too large. On the other hand,

$H \times SHE = TALK$. With $S = 9$ and $T = 1$, H must be 2, otherwise $H \times SHE$ cannot produce 1 in the first position. Now the divisor is either 923 or 927 and the quotient either 123 or 127. But 923×123 is a six digit number and hence not acceptable as a solution. If $E = 3$ is too large then so will be $E = 7$. So we have to go back and start with some other value for S.

Let $S = 6$. Then $E = 4$ or 6. We dismiss the $E = 6$ since 6 is already taken. Thus E has to be 4. Repeating the same arguments as above we get $T = 1$, and $H = 2$ or 3. Now, by fixing H, we know both divisor and quotient, so by multiplying them we can get the dividend. In the case $H = 2$, 624×124 leads $FEARS$ to be 77376 which is not acceptable because it makes $F = E = R$. The next case, $H = 3$, gives us $THE = 134$, $SHE = 634$ and $FEARS = 84956$. The puzzle is solved as follows:

$$
\begin{array}{r}
134 \\
634\,|\overline{84956} \\
634 \\
\overline{2155} \\
1902 \\
\overline{2536} \\
2536 \\
\hline
\end{array}
$$

The reader can check that $S = 1$ or 4 do not give valid solutions.

c) Answer: $GUM = 718$.

e) Answer: $CAN = 458$.

4.29. Right after 12 o'clock, the hands start moving. But since the minute hand goes faster, the hands will not coincide before the minute hand makes a complete cycle. Now the hour hand is at 1. They continue moving, now with the minute hand chasing the other hand from behind. After some time, before the minute hand makes another cycle, they must coincide again. At this moment the minute hand has made one complete cycle and a fraction of cycle, say λ. Meanwhile, the hour hand has only covered the fraction λ of one complete cycle. Since the minute hand moves 12 times faster,

$$1 + \lambda = 12\lambda$$

From the above we get $\lambda = \frac{1}{11}$ of a cycle. It takes one hour for the minute hand to complete one cycle, so after 1 hour and $\frac{60}{11}$ minutes the hands coincide again.

4.30. We have

$$
\begin{aligned}
1 &= f(1) \\
&= f\left(\frac{1}{2} + \frac{1}{2}\right) \\
&= f\left(\frac{1}{2}\right) + f\left(\frac{1}{2}\right) \\
&= 2f(\frac{1}{2})
\end{aligned}
$$

So $f(\frac{1}{2}) = \frac{1}{2}$.

4.31. Notice that $CRUDE = C \times 10000 + RUDE$. So we can cancel $RUDE$ from both sides of the equation and get

$$NUDE + NOT + NOR = C \times 10000.$$

On the other hand, the left hand side of the above equation is less than 12000. This forces C to be 1. Now we have

$$NUDE + NOT + NOR = 10000.$$

If N is 7 or less, then

$$
\begin{aligned}
NUDE + NOT + NOR &< 8000 + 800 + 800 \\
&= 9600 \\
&< 10000.
\end{aligned}
$$

Therefore, N must be 8 or greater. On the other hand, $N = 9$ is too large since

$$
\begin{aligned}
NUDE + NOT + NOR &> 9000 + 900 + 900 \\
&= 10800 \\
&> 10000.
\end{aligned}
$$

Hence $N = 8$.

Bearing this in mind, there are many solutions; among them are

$$8350 + 824 + 826 = 10000;$$
$$8251 + 873 + 876 = 10000;$$
$$8213 + 890 + 897 = 10000.$$

4.32. We have $4AYE = 3YES$. It follows that AYE is divisible by 3, and YES is divisible by 4. On the other hand, $4E$ and $3S$ should produce the same digit in their lowest position (first digit on the right). For example, if $E = 7$, then S has to be 6 so that both $4E$ and $3S$ would end with an 8. Because of this we get the following table.

If	E= 1	then	S= 8.
If	E= 2	then	S= 6.
If	E= 3	then	S= 4.
If	E= 4	then	S= 2.
If	E= 5	then	S= 0.
If	E= 6	then	S= 8.
If	E= 7	then	S= 6.
If	E= 8	then	S= 4.
If	E= 9	then	S= 2.
If	E= 0	then	S= 0.

We may dismiss the first five lines, because none of $Y18$, $Y26$, $Y34$, $Y42$, and $Y50$ could be divisible by four. We also dismiss $E = 0$ for the obvious reason that we do not want to have repeated digits.

Starting with $E = 6$ and $S = 8$, we observe that in this case AYE is even, and as a result YES has to be divisible by 8. So we have the following choices for YES.

$$YES = 168, 368, 568, 768, 968.$$

Now we have to check all these to find the ones which are valid solutions. It turns out that $YES = 768$ solves the problem with

$$567 + 567 + 567 + 567 = 768 + 768 + 768$$

The other cases do not give valid solutions.

4.33. Each hour the boats get $12 + 17 = 29$ miles closer. So each minute they get $29/60$ of a mile closer. So one minute before they collided they were $29/60$ of a mile apart.

4.34. Notice that $7 \cdot 11 \cdot 13 = 1001$, and multiplying a three digit number by 1001 is the same as writing it twice in a line.

4.35. For simplicity we assume that we have one unit of each liquid. We also assume that flask A contains water and flask B contains acid. After transferring liquids back and forth according to the problem, both flasks still hold one unit of mixed liquid each. Let r be the amount of water in flask B. Then flask B holds $1 - r$ units of acid. Since we started with one unit of acid, the rest of it, $1 - (1 - r) = r$, should be in flask A. So we have the same amount of acid in A that we have water in B.

4.36. Let us again, for the sake of simplicity, assume that we have one unit of each liquid. Let us also refer to the containers as in the previous problem. We also assume that each time we are transferring x units of liquid back and forth. The ratios of water and acid in flask A change only on even numbered pours and those of flask B change only on odd numbered pours. If both flasks are going to have the same amount of acid, then the ratio of acid in both flasks should be $1/2$. In flask B, the ratio at the beginning is 1 and each time that we pour the liquid back and forth the acid gets diluted and the ratio decreases. Is it the case that it will eventually become $1/2$ after finitely many pours?

If that is the case, there will be a first time that this happens. Just before that, the ratio of acid in both flasks is different from $1/2$. Assume there is r units of acid per 1 unit of liquid in flask B. So in the other flask there is $1 - r$ units of acid per one unit of liquid. If we take x units of this mixture add them to flask B and mix the content, the new ratio of acid, R, satisfies the following

$$R = \frac{r + (1 - r)x}{1 + x}. \tag{4.2}$$

We are interested in the conditions that the new ratio, R, becomes

1/2. Substituting 1/2 for R in (4.2), we have the following

$$\frac{1}{2} = \frac{r + (1-r)x}{1+x}.$$

Thus,

$$1 + x = 2r + 2x - 2rx.$$

Finally,

$$(2r - 1)(1 - x) = 0$$

The only solutions to the above equation are $x = 1$ or $r = 1/2$. It follows that unless each time we empty the whole contents of each flask into the other or that the ratio is already 1/2, the new ratio will not be 1/2. None of these are the case, so the answer to the question is "no".

However, if we subtract 1/2 from R (the new ratio) and factor the result we get

$$R - \frac{1}{2} = \left(\frac{1-x}{1+x}\right)\left(r - \frac{1}{2}\right)$$

This means that each time the difference is multiplied by the factor $(1-x)/(1+x)$ which is less than 1. So although we never reach 1/2 we can get as close as we want.

Exercise: Prove that, after $2n$ pours $r = \dfrac{1}{2} + \dfrac{1}{2}\left(\dfrac{1-x}{1+x}\right)^n$.

4.37. We cannot cut the cube into 27 sub-cubes with fewer than six cuts. Suppose that we have cut the cube into 27 sub-cubes of equal size some way or another. Imagine that they are still in their original position as part of the larger cube. One of these sub-cubes is completely in the center of the larger cube while the rest of them have at least one face which is part of one of the faces of the original cube. All the faces of the inner cube have to be cut. It has six faces so we need at least six planar cuts.

4.38. Let us refer to the three men by B, M and F, where B refers to the man in the back row, M to the man in the middle and F to the man in the front.

Case 1: F and *M* have black hats. Then *B*, knowing that there are only two black hats, concludes that he, himself must have a red hat and declares this. Now *M* argues that the only way for *B* to be able to tell the color of his hat would be that he sees two black hats. So he, *M*, concludes that his own hat must be black. In the same way *F* concludes that his hat must be black too.

Case 2: F has a black hat and *M* has red hat. Then *B* declares that he cannot determine the color of his hat. Now *M* knows that either *F* or himself, must have a red hat. Since he sees that *F* has a black hat he concludes that his own hat should be red. So he answers "red" to the question. Likewise *F* knows that *B* must have seen at least one red hat in the two front rows. But if both *F* and *M* had red hats then *M* would not have been able to tell the color of his hat. Therefore *F* declares that his hat is black.

Case 3: F has a red hat. Now *B* cannot tell the color of his hat. But this is not new information for *M* since he sees that *F* has a red hat. He himself could have either color. So he too cannot tell the color of his hat. *F* knows that in the two front rows there is at least one red hat. On the other hand, if he had a black hat then *M* would have been able to determine the color of his own hat. Since this is not the case *F* knows that his own hat must be red.

So regardless of how the hats are distributed, the front man who does not see any of the hats is the only one who can always determine the color of his hat.

4.39. The assumption that at both times scientists were expressing the truth is wrong, and we can conclude anything from a wrong assumption.

4.40. The fact the Erdös number of the author of this book is 1 means that he has written a paper with somebody whose Erdös number is zero. There is only one such person and that is Erdös himself. So the author of this book has written a paper with Erdös.

4.41. Let us color the chess board in the following way.

If, in fact, the knight could visit all the squares exactly once and come back to the original square, it should not matter from which square it starts. If such a loop exists starting from one square, it will exist if we start from any other square.

Suppose that the knight starts from the lower left corner of the chess board. This square is colored white. The knight has only two squares to which it can move, so if we are able to complete a loop, we have to come back to the original square via one of these two. We also need one of them to get out of the corner square. Notice that both of these squares are colored white. So the first move and the last move are to and from a white square. In the mean time, we have to visit all the black squares exactly once. But notice that the only time the knight can go to a black square from a white square or vice versa is in the two middle rows. Otherwise the knight cannot change color. Since the first and last moves are from white squares, the first black square and the last black square that the knight visits both have to be in the two middle rows.

There are 8 black squares in the lowest and highest rows. Each time that the knight is in one of these squares, the next move will be to a black square in the middle two rows (these are the only moves available). So in order to visit all the black squares in the top and bottom rows, we need a total of nine black squares in the middle two rows: one to start with, and one after each visit to one of the 8 squares in the top and bottom rows. But we only have 8 black squares in the middle two rows. Hence, the answer to the question is no.

But it is possible to cover all the squares exactly once. We leave it to the reader to find a solution. Only remember that in this case the starting position is important.

4.42. Suppose that the squares are colored alternately in black and white, just like a standard chess board.

The moves of the knight will alternate from black squares to white. The entire tour will require 49 moves. Therefore if the tour begins on a black square then it will end on a white. If it begins on a white square then it will end on a black. In conclusion, the tour cannot begin and end on the same square.

4.43. One can ask her if she is a woman. If she answers with left to right motion of head, we know she is from that particular tribe. Any other motion of head tell us that she is not from that tribe.

4.44. Since the villain has heard of the other two actors and Curly has not heard of Bob, Curly is not the villain. So he is either the loonie or the leading man. If Bob were the loonie then Curly had to be the leading man but then he must have heard of Bob. So Bob is not the loonie. Likewise he is not the leading man. Then he must be the villain. On the other hand Bob makes more money than Joe, so Joe is not the loonie who we know earns much more than the villain, Bob. Hence Joe is the leading man and Curly is the loonie.

4.47. We need some notation before we start to solve the problem. Let A, B, C and D denote the four players of the game and let

$$P(X) = \mathbf{Pr}\{X \text{ gets a Yarborough hand}\},$$
$$P(X, Y) = \mathbf{Pr}\{X \text{ and } Y \text{ get Yarborough hands}\}.$$

Also, let P denote the probability that at least one player gets a

Yarborough hand. It would seems reasonable that

$$P = P(A) + P(B) + P(C) + P(D),$$

but this equation needs some corrections. If we examine the right hand side of the above equality, we see that $P(A, B)$ has been included twice, both in $P(A)$ and in $P(B)$. The same is true for $P(B, C)$ and so on. To correct this, we have to subtract all the combinations of $P(X, Y)$ once. Now we have

$$P = P(A) + P(B) + P(C) + P(D) - P(A, B) - \cdots - P(C, D).$$

Since no more than two players can get Yarborough hands at the same time, no other terms are needed and the above equation is correct. Notice that all $P(X)$'s are equal. Similarly all different combinations of $P(X, Y)$ are equal. Thus we can write

$$P = 4P(A) - 6P(A, B). \tag{4.3}$$

There are $\binom{36}{13}$ possible Yarborough hands out of $\binom{52}{13}$ possible hands so

$$P(A) = \frac{\binom{36}{13}}{\binom{52}{13}}.$$

If A has a Yarborough hand then there are $\binom{23}{13}$ Yarborough hands out of the $\binom{39}{13}$ left for B. So

$$P(A, B) = \frac{\binom{36}{13}\binom{23}{13}}{\binom{52}{13}\binom{39}{13}}.$$

Substituting these into (4.3), we get

$$P = 4\frac{\binom{36}{13}}{\binom{52}{13}} - 6\frac{\binom{36}{13}\binom{23}{13}}{\binom{52}{13}\binom{39}{13}} \approx 0.0145528.$$

4.48. In a fair game of roulette, if a player's number comes up, she will win 36 times the money she had bet, because the odds of winning are 1 in 36. In the same way, if she guesses the color right, she should be paid twice her bet. In this game if someone plays the game over and over the average gain or loss should be zero, whereas in the new game, for each 38 dollars which one spends, on average one can only win 36 dollars. So one loses $2 out of each $38. It is perhaps clearer to say that the house wins 2 dollars out of each 38 dollars which are bet in the game.

4.49. Let us answer the last question first. To win $40 or more, the winning number should at least have the first three digits or the last three digits in common with 987654. There are 1000 integers of the form $XXX654$ and 1000 integers of the form $987XXX$. Since 987654 is counted twice, and it is the only number which is counted twice, there are $1000 + 1000 - 1 = 1999$ integers of either form. So the chances that the winning number be one of these are 1999 out of 1000000.

Similarly, we can calculate the odds of winning $200 or more, $2000 or more and so on. The following table shows the result.

Odds of winning	$50,000	or more are	1 in 900,000
Odds of winning	$2,000	or more are	19 in 900,000
Odds of winning	$200	or more are	199 in 900,000
Odds of winning	$40	or more are	1999 in 900,000

We get another table which shows the odds of winning exactly each prize. For example to calculate the odds of winning $40, we have to exclude the cases for which we would win $200 or more. Hence for $1899 - 189$ cases out of 900000 we win exactly $40. The rest of the table is calculated the same way.

Odds of winning	50,000	are	1 in 900,000
Odds of winning	2,000	are	18 in 900,000
Odds of winning	200	are	180 in 900,000
Odds of winning	40	are	1800 in 900,000

If we compare the first two rows of the above table, we see that the odds of winning are 18 times better in the second row but the payoff is 25 times less, whereas it should have been 18 times less if both payoffs were equally fair. The payoffs in the second and the third row are equally fair.

Notice that none of the payoffs is fair according to the odds.

Chapter 5

Recreational Math

5.1. Adding all the squares from 1 to 27 we obtain $6930 = 3 \cdot 2310$. Thus, the weight of each subset must be 2310. One possible solution, obtained using a computer, is

$$
\begin{aligned}
2310 &= 1^2 + 2^2 + 3^2 + 4^2 + 5^2 + 6^2 + 7^2 + 8^2 + 9^2 + 10^2 \\
&\qquad + 11^2 + 13^2 + 14^2 + 15^2 + 18^2 + 19^2 + 23^2 \\
2310 &= 12^2 + 17^2 + 24^2 + 25^2 + 26^2 \\
2310 &= 16^2 + 20^2 + 21^2 + 22^2 + 27^2.
\end{aligned}
$$

5.3. A possible strategy is to carry as many gallons as needed in five 100-mile steps. Note that the jeep can go from the starting point to the first step and back without refueling. Thus, for each step, the jeep needs 4 cans of gas in order to deliver 3 (the extra can is used as fuel for the jeep). Deliver first a certain amount of gas to the first step. Note that once all the gas has been delivered to this first step, the tank of the jeep is still half full. Fill it up (so now we have one container with only 5 gallons of gas) and repeat the same process to get to step 2, and so on. One can count carefully how many gallons are needed so that exactly 100 are delivered at the last step: we will have 430 at the start (assuming that the jeep's tank is empty at first; if it is full, we need only 420), 320 at the first step, 235 at the second, 180 at the third, 135 at the fourth and 100 at the last step.

5.6. One should always switch. Note that the first time one chooses an envelope, one has the same probability of choosing the one with more money as the one with less money. This means that, once one has chosen one envelope, the other envelope has $450 with probability 1/2 and $50 with probability 1/2 also. Thus the average gain is

$$\frac{\$450}{2} + \frac{\$50}{2} = \$250 > \$150.$$

If one envelope has twice as much as the other, the average gain is $187.50, and if one envelope has one and a half times the amount in the other, the average gain is $162.50. In any case, one should always switch.

(A good way to understand why it is always good to switch is to suppose that one envelope has, say, one-hundred times as much as the other. Then the other envelope would have either $1.50 or $15,000 (with probability 1/2 (!)). It seems clear that one should switch.)

5.8. If we denote the checkers in their initial positions by the numbers 1, 2, 3, 4, 5, 6, 7, 8, 9, 10, then we can proceed as follows. Put 7 on 10, then 5 on 2, then 3 on 8, then 1 on 4, and lastly 9 on 6. Thus they are arranged in pairs on the places originally occupied by the counters 2, 4, 6, 8, 10.

5.9. Take any conventional 3×3-magic square, multiply all its entries by 2 and subtract 1 from all of them. You obtain a magic square whose entries are the first nine odd integers.

5.10. Take any conventional 3×3-magic square, multiply all its entries by 2. You obtain a magic square whose entries are the first nine even integers.

5.11. Let $a + 1, a + 2, \ldots, a + 9$ be any sequence of nine consecutive integers. Take any conventional 3×3-magic square, and add a to all its entries. You obtain a magic square whose entries are $a + 1, a + 2, \ldots, a + 9$, as desired.

5.12. Put three pearls in each tray. One of them will not balance, so the odd pearl is among the three in this tray. Now weigh these three pearls—one in each tray. One will not balance, and that is the odd one.

5.13. With twelve, put four pearls in each tray. One of them will not balance, so the odd pearl is among the four in this tray. Now weigh three of these four pearls, and keep the fourth in your hand. If one tray does not balance, it contains the odd one. If the trays balance, then the odd one is the one you have in your hand.

With fifteen, put also four pearls in each tray. If it balances, the odd pearl is among the remaining three, which can be weighted to find the odd one. If it does not balance, just proceed as with twelve.

5.14. Knowing if the pearls are heavier or lighter does not make any difference in the previous two exercises. Note that, since the scale compares three groups, it shows whether the odd pearl is heavier or lighter.

With 1 weighing we can handle up to 4 pearls: weigh 3 of them; if they balance, the odd pearl is the one we left out; if they do not, the balance will show that 2 pearls have the same weight and the other one is heavier or lighter.

With 2 weighings, we can handle up to 16 as follows: divide them into 4 groups of 4 pearls each and weigh three of these groups, so that we know which one of the 4 groups contains the odd pearl. Then proceed as in the discussion for the 1-weighing case.

With 3 weighings, we can handle up to $16 \times 4 = 64$ as follows: divide them into 4 groups of 16 pearls each and weigh three of these groups, so that we know which one of the 4 groups of 16 pearls contains the odd pearl. Then proceed as in the discussion for the 2-weighings case.

5.15. The only 2×2 Latin squares are

$$\begin{matrix} a & b \\ b & a \end{matrix} \quad \text{and} \quad \begin{matrix} b & a \\ a & b \end{matrix},$$

where a and b are arbitrary objects.

There are twelve 3×3 Latin squares, namely all the squares generated by all the permutations of $\{a, b, c\}$ in

$$
\begin{array}{ccc}
a & b & c \\
c & a & b \\
b & c & a
\end{array}
\qquad \text{and} \qquad
\begin{array}{ccc}
a & b & c \\
b & c & a \\
c & a & b.
\end{array}
$$

Note that given, without loss of generality, that the first row is a, b, c, then in the second row we have only two choices for a, and the positions of b and c are forced once a is located. So only two cases are possible up to permutations of $\{a, b, c\}$.

To find an upper bound to the number of 8×8 Latin squares, we can do the following. There are $8!$ arrangements of the first row (all the permutations of the 8 objects). Given one such arrangement, there are $7!$ arrangements of the first column (the permutations of 7 objects, since the object that occupies the top left corner is already fixed). Then there are $7!$ ways to arrange the second row, once we have an arrangement of the first row and column, and then there are $6!$ arrangements of the second row. Continuing in this fashion we find that an upper bound to the number of 8×8 Latin squares is

$$
8!\,7!\,7!\,6!\,6!\,5!\,5!\,4!\,4!\,3!\,3!\,2!\,2! = 63,415,300,800,997,490,688 \cdot 10^7
$$

This bound is actually very rough. The actual number of 8×8 Latin squares is $108,776,032,459,082,956,800$, a much smaller number, but this fact is very hard to prove.

One can easily derive a lower bound. Let us count the possibilities per row. Denote the objects by a, b, c, d, e, f, g, h. For the first row, we clearly have $8!$ arrangements (all the ways to permute 8 elements). Let us see in how many ways we can arrange the second row for each given first row. The object a can be placed in 7 positions in the second row (all but the one that it occupied in the first row). For each position that a takes, the object b can be placed at least in 6 positions (all but the one that it occupied in the first row and the one that is already taken by a). Note that, if a in the second row is placed in the same column as b in the first row, then we can put b in 7 different positions in the second column, so 6 is just a lower bound. For each position that a and b take, the object c can be placed in at least 5 positions (all but the one that it occupied in the first row and the ones that are already taken by a and b). As before, there are cases where c can actually take 6 different positions. Following in this fashion, we see that, for each arrangement of the first row, there are at least 7! arrangements of the second row. We can reason in the same manner for rows 3 to 8, and obtain at least 6! arrangements of the third row for each arrangement of the first and second row, at least 5! arrangements of the fourth row for each arrangement of the first, second and third row, etc. Thus, at least we have

$$8!7!6!5!4!3!2! = 5,056,584,744,960,000$$

different 8×8 Latin squares.

For more on Latin Squares, see [BAL, p. 189 ff.].

5.16. The next elements are 1 (or 001), 121, 441, 961, 691, 522, ... (Hint: read the numbers backwards.)

5.17. A configuration that produces a periodic population is

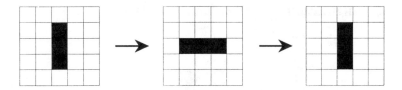

Unless we have infinitely many people, a static configuration is impossible. If there are only one or two squares, they die. If there are more, one can easily check that if one person has three neighbors, an offspring is always produced. Thus, for the configuration to remain static, each person must have at most two neighbors. On the other hand, the following configurations also produce offsprings:

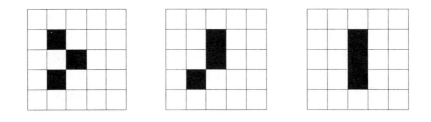

Thus, the only possibility is if all the squares are arranged in a diagonal line:

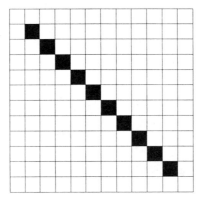

But then the ones at the ends will die of loneliness (of course, unless the line has no ends, for which we need infinitely many people).

A configuration that dies immediately is

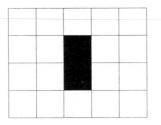

There are configurations that grow without bound. One such configuration was found in 1970 by B.W Gosper. It is called the 'Glider Gun'. It looks as follows:

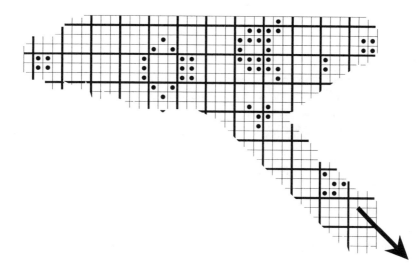

The reader can check that every 30 generations, a new 'glide' is shot from the gun (top of the figure) in the southeast direction. For more on 'The Game of Life' and other games, we refer the interested reader to the book *Winning Ways,* by Elwyn R. Berlekamp, John H. Conway and Richard K. Guy.

5.18. The probability that a six does not come up in the first 30 rolls is

$$p = \left(\frac{5}{6}\right)^{30} \approx 0.00421272.$$

The game will not be unfair for you if

$$p \geq \frac{L}{L + W},$$

where L is the amount you pay if you lose and W is the amount you receive if you win (see **Chapter 3** for details). Thus we should play the game if and only if we have the inequality

$$\left(\frac{5}{6}\right)^{30} \geq \frac{100}{1,000,100}.$$

A calculator can be used to show that this inequality does hold. Therefore you should play the game.

5.19. Take the goat to the other side, come back, pick up the wolf, take it to the other side and bring the goat back to its original position. Then pick up the cabbages, take them to the other side and leave them with the wolf. Finally, come back and pick up the goat. Bring the goat to the other side, and you are done.

5.20. Fifteen trips are required. Start by taking two boys to the other side, then let one boy return with the boat; then bring one man over to the other side, letting the other boy return with the boat. At this point we have one man on the other side and two men and all three boys at the starting point, and so far we have made four trips.

Repeat this same process two more times, so that after a total of 12 trips all three men are on the other side and all three boys are at the starting point. Then bring two boys to the other side, let one return with the boat and bring the last boy over to the final destination. We have made a total of fifteen trips.

Note that at each stage the strategy is obvious. There is only one variation: in the process above, after the first two boys went to the other side and one came back, he could also take the third boy to the other side (instead of letting a man cross) and then come back with the boat. This procedure gives the same total number of trips.

5.21. We can put a maximum of 16 kings. Note that we cannot put more than 4 kings per row. If we have 4 kings in a row the next row must be empty, and if we have 3 kings, the next one can only have 1 king. Therefore we cannot put more than 16 kings. A possible configuration with 16 kings is the following:

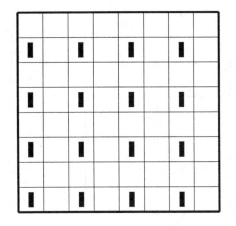

5.22. We need at least five queens. The following is a possible configuration:

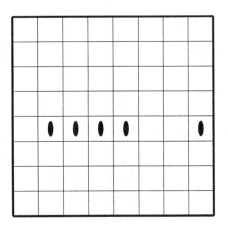

For more in chessboard recreations, see [BAL, Ch. 6].

5.23. Note that all the vertices have an even number of edges except for two of the central ones, which have five. Thus we have to start at one of these vertices with five edges and finish at the other one. A possible route is shown below; the numbers denote the sequence of paths.

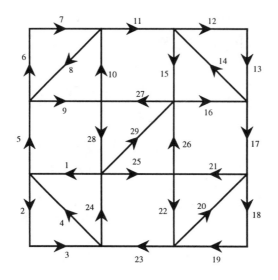

5.24. Proceed as in similar exercises in Chapter 3. The answer is 1, the first digit of the number 10,222.

5.25. *Step 1:* Fill the 5 oz. and the 11 oz. containers with the liquid. Note that exactly $24 - (5 + 11) = 8$ oz. remain in the original container.

Step 2: Pour all the contents of the 5 oz. container in the 13 oz. container, and refill the rest with the contents of the 11 oz. container. Note that exactly $11 - (13 - 5) = 3$ oz. remain in the 11 oz. container.

Step 3: Fill the 5 oz. container with the contents of the 13 oz. container, and then empty the 5 oz. container in the 11 oz. container. Now we have 8 oz. of liquid in the original container, 8 oz. of liquid in the 11 oz. container and 8 oz. of liquid in the 13 oz. container .

5.26. There is no general method to do this kind of problem. In this case we have used a computer mini-program to find all the possible combinations, and then we choose the most economical. It is easy to find a route that will cost $1,650, namely SF, NY, Ch, Tu, Da, De, SF. So we can put this number as an upper bound and make the program find those routes that cost less than $1,650. Using

Mathematica, the following program gives the answer in about 4 seconds:

```
A=Permutations[{2,3,4,1,6}];

B={{0,300,300,225,400,275}, {300,0,350,400,375,250},
    {300,350,0,250,325,250}, {225,400,250,0,225,275},
    {400,375,325,225,0,400},{275,250,250,275,400,0}};

CC={Ch, Da, De, NY, SF, Tu};

T[n_]:={B[[5,A[[n,1]]]], B[[A[[n,1]], A[[n,2]]]],
    B[[A[[n,2]], A[[n,3]]]], B[[A[[n,3]], A[[n,4]]]],
    B[[A[[n,4]], A[[n,5]]]], B[[A[[n,5]],5]]};

TT[n_]:=Apply[Plus, T[n]];

For[n=1, n<121, n++,
    If[TT[n] < 1650,
        Print[{SF, CC[[A[[n,1]]]], CC[[A[[n,2]]]],
                CC[[A[[n,3]]]], CC[[A[[n,4]]]],
                        CC[[A[[n,5]]]], SF},
                    T[n]," ", TT[n]
            ]
        ]
    ]
```

Output:

```
{SF, Da, Tu, De, Ch, NY, SF}{375, 250, 250, 300, 225, 225} 1625
{SF, De, Tu, Da, Ch, NY, SF}{325, 250, 250, 300, 225, 225} 1575
{SF, NY, Ch, Da, Tu, De, SF}{225, 225, 300, 250, 250, 325} 1575
{SF, NY, Ch, De, Tu, Da, SF}{225, 225, 300, 250, 250, 375} 1625
```

Thus, the most cost-efficient route is SF, De, Tu, Da, Ch, NY, SF, (of course in either direction), and the total cost is \$1,575.

5.27. Let us denote the elements of the matrix by a_{ij}, $1 \leq i \leq k$, $1 \leq j \leq m$. Let r_i be the product of all the entries in row i, and c_j the product of all the entries in column j. That is,

$$r_i = a_{i1} \cdot a_{i2} \cdots a_{im} \quad \text{and} \quad c_j = a_{1j} \cdot a_{2j} \cdots a_{kj}.$$

Note that

$$
\begin{aligned}
c_1 \cdot c_2 & \cdots c_m \cdot r_1 \cdot r_2 \cdots r_k \\
&= \left(a_{11} \cdot a_{21} \cdots a_{k1}\right) \cdot \left(a_{12} \cdot a_{22} \cdots a_{k2}\right) \cdots \left(a_{1m} \cdot a_{2m} \cdots a_{km}\right) \\
&\quad \cdot \left(a_{11} \cdot a_{12} \cdots a_{1m}\right) \cdot \left(a_{21} \cdot a_{22} \cdots a_{2m}\right) \cdots \left(a_{k1} \cdot a_{k2} \cdots a_{km}\right) \\
&= \prod_{i=1}^{k} \prod_{j=1}^{m} a_{ij}^2 \\
&= 1
\end{aligned}
$$

Therefore, if we want c_j and r_i to be -1 for all i, j, we must have that $k + m$ is an even number. Thus, for $k + m$ odd, there are no $k \times m$ matrices of this type.

When $k + m$ is even we can do the following. First note that, for each row, we have 2^m possibilities (since for each entry we have 2 possibilities and we have m entries). This is without taking into consideration that we want the product of its entries (r_i) to be -1. For how many of these do we have $r_i = -1$? Well, for exactly half of them (this may take a minute of thought, but note that for each combination in which $r_i = 1$, we have another for which $r_i = -1$—just multiply all the entries by -1. Thus the combinations with $r_i = 1$ are in one-to-one correspondence with those for which $r_i = -1$). Therefore, we have $2^m/2 = 2^{m-1}$ possibilities for each row.

We can now write the first $k - 1$ rows arbitrarily, and we have $(2^{m-1})^{k-1}$ overall possibilities. Now note that in the last row there are no choices since we need to make sure that $c_j = -1$ in each column. So far we have $c_j = -1$ for $j = 1, 2, \ldots m$ and $r_i = -1$ for $i = 1, 2, \ldots k - 1$. It only remains to check that $r_k = -1$. By the formula above, and using the fact that if $k + m$ is even then

$k + m - 1$ is odd, we have

$$
\begin{aligned}
c_1 \cdot c_2 \cdots c_k \cdot r_1 \cdot r_2 \cdots r_m &= (-1)^{k+m-1} \cdot r_k \\
&= -r_k \\
&= 1.
\end{aligned}
$$

Therefore we must also have $r_k = -1$. Thus, the total number of $k \times m$ matrices of this type is

$$
\begin{aligned}
2^{(k-1)(m-1)} \qquad &\text{for } k + m \text{ even} \\
0 \qquad &\text{for } k + m \text{ odd.}
\end{aligned}
$$

Chapter 6

Algebra and Analysis

6.1. We can proceed as follows:

$$(1-a)(1-b)(1-c)(1-d)$$
$$= 1 - a - b - c - d + ab + ac + ad + bc + bd + cd$$
$$- abc - abd - acd - bcd + abcd$$
$$= 1 - a - b - c - d + ab(1-c) + bc(1-d)$$
$$+ cd(1-a) + ad(1-b) + ac + bd + abcd$$
$$\geq 1 - a - b - c - d,$$

since $0 \leq a, b, c, d \leq 1$ implies

$$ab(1-c) + bc(1-d) + cd(1-a) + ad(1-b) + ac + bd + abcd$$
$$\geq 0.$$

6.2. Note that, for any a, $(a-1)^2 \geq 0$, which implies $a^2 + 1 \geq 2a$. Therefore we have, for positive a,

$$\frac{a^2 + 1}{a} \geq 2.$$

Since the same holds for b, c and d, we have

$$\frac{(a^2+1)(b^2+1)(c^2+1)(d^2+1)}{abcd} = \frac{a^2+1}{a} \cdot \frac{b^2+1}{b} \cdot \frac{c^2+1}{c} \cdot \frac{d^2+1}{d}$$
$$\geq 2 \cdot 2 \cdot 2 \cdot 2$$
$$= 16.$$

6.3. Let

$$M = 1 + \frac{1}{2} + \frac{1}{3} + \cdots + \frac{1}{n},$$

with $n > 1$. Let k be the greatest integer such that $2^k \leq n$. Note that $k \geq 1$. By the definition of k, we must have

$$2^k \leq n < 2^{k+1}.$$

Now note that none of the numbers $2^k, 2^k + 1, 2^k + 2, \ldots, n$ is divisible by 2^k (if one of these numbers is divisible by 2^k, then it must be of the form $\ell \cdot 2^k$, with $\ell \geq 2$. But then we would have $2 \cdot 2^k \leq n$, contrary to the choice of k). Thus, only one of the denominators in the right hand side of

$$M = 1 + \frac{1}{2} + \frac{1}{3} + \cdots + \frac{1}{n}$$

is divisible by 2^k. Now, let D be the least common multiple of $1, 2, 3, \ldots n$. Note that D is divisible by 2^k, but not divisible by 2^{k+1}. Then, writing the expression above with a common denominator, we have

$$
\begin{aligned}
M &= 1 + \frac{1}{2} + \frac{1}{3} + \cdots + \frac{1}{n} \\
&= \frac{D + D/2 + D/3 + \cdots + D/2^k + \cdots + D/n}{D}.
\end{aligned}
$$

Finally, note that every summand in the numerator of the last expression is even except for $D/2^k$, which is odd. Thus, the numerator must be odd (since the sum of even numbers is even and the sum of an even number plus an odd number is odd), whereas the denominator is even (since D is divisible by 2^k). But an odd number divided by an even number is never an integer. So M cannot be an integer.

6.4. First let us prove that $n^7 - n$ is always divisible by 7 (this is **CHALLENGE PROBLEM 6.1.3**). Now, $n^7 - n$ factors as

$$(n-1)n(1+n)(n^2 - n + 1)(n^2 + n + 1).$$

If n is of the form $7k + 1$, $7k$ or $7k - 1$ then the result is clear. If n is of the form $7k + 2$ then

$$n^2 + n + 1 = (7k + 2)^2 + (7k + 2) + 1 = 1 + 2 + 4 + 7 \cdot \text{(integer)},$$

which is divisible by seven. The same thing happens if n is of the form $7k - 3$.

If n is of the form $7k - 2$ then

$$n^2 - n + 1 = (7k - 2)^2 - (7k - 2) + 1 = 1 + 2 + 4 + 7 \cdot \text{(integer)},$$

which is divisible by seven. The same thing happens if n is of the form $7k + 3$. Therefore $n^7 - n$ must always be divisible by 7. This also proves that if n is not a multiple of 7, then $n^6 - 1$ is divisible by seven. In other words, $n^6 = 7K + 1$, for K an integer. Using the binomial theorem, proved in the text, we can see that

$$n^{6\ell} = (7K + 1)^\ell = 1 + 7L,$$

for some integer L.

After these preliminaries we can easily solve the exercise:

$2222^{5555} + 5555^{2222}$ can be written as

$$(7 \cdot \text{(integer)} + 3)^{6 \cdot \text{(integer)}+2} + (7 \cdot \text{(integer)} + 4)^{6 \cdot \text{(integer)}+5},$$

which, using the binomial theorem, can be written as

$$3^{6 \cdot \text{(integer)}+2} + 4^{6 \cdot \text{(integer)}+5} + 7 \cdot \text{(integer)}.$$

Thus we only have to show that

$$3^{6 \cdot \text{(integer)}+2} + 4^{6 \cdot \text{(integer)}+5}$$

is divisible by 7. Using $n^{6\ell} = 1 + 7 \cdot \text{(integer)}$, we can write the last expression as

$$3^2 + 4^5 + 7 \cdot \text{(integer)} = 1,033 + 7 \cdot \text{(integer)},$$

which is divisible by 7.

6.5. We can use the binomial theorem as follows:

$$11^{10} - 1$$
$$= (10 + 1)^{10} - 1$$
$$= \left[\binom{10}{0} 10^{10} + \binom{10}{1} 10^{9} + \cdots + \binom{10}{8} \cdot 100 \right.$$
$$\left. + \binom{10}{9} \cdot 10 + 1 \right] - 1$$
$$= \left[\binom{10}{0} 10^{10} + \binom{10}{1} 10^{9} + \cdots + \binom{10}{8} \cdot 100 + 10 \cdot 10 \right]$$
$$= 100 \left[\binom{10}{0} 10^{8} + \binom{10}{1} 10^{7} + \cdots + \binom{10}{8} + 1 \right].$$

6.6. Let N be a positive integer. Consider the set

$$\{1, 10, 100, 1000, \ldots, 10^{N \cdot N}\}.$$

This set contains $N^2 + 1$ elements. On the other hand every element in this set can be written in the form

$$10^k = c_k \cdot N + r_k,$$

where $0 \le r_k < N$. By the Pigeonhole Principle, since we have $(N^2 + 1)$ $r'_k s$ and only $N - 1$ possible values for each r_k, there must be at least one r_k (call it r_0) that is repeated N times. That is, there are N positive integers $k_i \le N^2 + 1$ such that

$$10^{k_i} = c_{k_i} \cdot N + r_0, \quad i = 1, 2, 3, \ldots N.$$

Adding all these numbers we obtain

$$10^{k_1} + 10^{k_2} + \cdots + 10^{k_N}$$
$$= N \cdot (c_{k_1} + c_{k_2} + \ldots + c_{k_N}) + \underbrace{(r_0 + r_0 + \cdots + r_0)}_{N \text{ times}}$$
$$= N \cdot (c_{k_1} + c_{k_2} + \ldots + c_{k_N}) + N \cdot r_0.$$

Thus, $10^{k_1} + 10^{k_2} + \cdots + 10^{k_N}$ is divisible by N, and its only digits are 0 and 1 (note that all the $k'_i s$ are different).

To show that, if N is not divisible by 2 or 5, then N divides a number whose digits are only 1's, we use a similar argument. Let

$$A_k = \underbrace{1111\cdots 1}_{k \text{ times}}, \quad k = 1, 2, 3, \ldots N.$$

Then, as before, each A_k can be written as

$$A_k = c_k \cdot N + r_k,$$

with $0 \leq r_k < N$. Since there are N numbers $r_k's$ and $N - 1$ possible values for r_k, by the Pigeonhole Principle there must be at least one r_k (call it r_0) that is repeated twice. That is, there are two numbers k_1, k_2 between 1 and N such that

$$A_{k_1} = c_{k_1} \cdot N + r_0$$

and

$$A_{k_2} = c_{k_2} \cdot N + r_0.$$

Without loss of generality, assume that $k_1 > k_2$. We have

$$A_{k_1} - A_{k_2} = N \cdot (c_{k_1} - c_{k_2}).$$

Thus N divides $A_{k_1} - A_{k_2}$. On the other hand, $A_{k_1} - A_{k_2}$ has the form

$$A_{k_1} - A_{k_2} = \underbrace{1111\cdots 1}_{(k_1-k_2) \text{ times}}\underbrace{0000\cdots 0}_{k_2 \text{ times}}.$$

Thus we have that N divides the number

$$2^{k_2} \cdot 5^{k_2} \cdot \underbrace{1111\cdots 1}_{(k_1-k_2) \text{ times}}.$$

But since N and 2 are relatively prime and N and 5 are also relatively prime, N cannot divide $2^{k_2} \cdot 5^{k_2}$. Therefore, N divides the number

$$\underbrace{1111\cdots 1}_{(k_1-k_2) \text{ times}}.$$

(Note that the argument in the second part of the exercise would also work for the first part, since we first proved that any number N has a multiple of the form $\underbrace{1111\cdots 1}_{(k_1-k_2) \text{ times}}\underbrace{0000\cdots 0}_{k_2 \text{ times}}$.)

6.7. It is easy to check with a pocket calculator that

$$\left(\frac{99}{101}\right)^{50} + \left(\frac{100}{101}\right)^{50} < 1 = \left(\frac{101}{101}\right)^{50}.$$

Therefore, since

$$\left(\frac{99}{101}\right)^{N} \le \left(\frac{99}{101}\right)^{50}$$

and

$$\left(\frac{100}{101}\right)^{N} \le \left(\frac{100}{101}\right)^{50}$$

for any $N \ge 50$, we must have

$$\left(\frac{99}{101}\right)^{N} + \left(\frac{100}{101}\right)^{N} < 1 = \left(\frac{101}{101}\right)^{N}.$$

or, multiplying by 101^{N},

$$99^{N} + 100^{N} < 101^{N},$$

in particular, for all $N \ge 1000$.

6.8. The formula is clearly true for $k = 1$. Assume that the formula is true for $k = N$, i.e. that

$$(a_1 + a_2 + \cdots + a_N) \cdot \left(\frac{1}{a_1} + \frac{1}{a_2} + \cdots + \frac{1}{a_N}\right) \ge N^2$$

holds for any positive real numbers $a_1, a_2, \ldots, a_n$. We want to prove that

$$(a_1 + a_2 + \cdots + a_{N+1}) \cdot \left(\frac{1}{a_1} + \frac{1}{a_2} + \cdots + \frac{1}{a_{N+1}}\right) \ge (N+1)^2.$$

Without loss of generality, assume that a_1 is the greatest of all the a_j's. We have:

$$(a_1 + a_2 + \cdots + a_{N+1}) \cdot \left(\frac{1}{a_1} + \frac{1}{a_2} + \cdots + \frac{1}{a_{N+1}}\right)$$

$$= (a_1 + [a_2 + \cdots + a_{N+1}]) \cdot \left(\frac{1}{a_1} + \left[\frac{1}{a_2} + \cdots + \frac{1}{a_{N+1}} \right] \right)$$

$$= 1 + (a_2 + \cdots + a_{N+1}) \left(\frac{1}{a_2} + \cdots + \frac{1}{a_{N+1}} \right)$$

$$+ \left(\frac{a_1}{a_2} + \frac{a_2}{a_1} \right) + \left(\frac{a_1}{a_3} + \frac{a_3}{a_1} \right) + \cdots + \left(\frac{a_1}{a_{N+1}} + \frac{a_{N+1}}{a_1} \right)$$

$$\geq 1 + N^2 + 2(N+1)$$

$$= (N+1)^2,$$

where we used the fact that $a_1/a_j + a_j/a_1 \geq 2$ (which can be easily proved using **PROBLEM 6.2.1** in the text). Note that the inductive hypothesis was used in the inequality above (we used

$$(a_2 + a_3 + \cdots + a_{N+1}) \cdot \left(\frac{1}{a_2} + \frac{1}{a_3} + \cdots + \frac{1}{a_{N+1}} \right) \geq N^2).$$

6.9. We can count them as follows: We have:

1 for every 10 numbers, in the units slot .... 10^7
10 for every 100 numbers, in the tens slot .... 10^7

$$\vdots$$

10^6 for every 10^7 numbers, in the millions slot .... 10^7

TOTAL ....$7 \cdot 10^7$

6.10. We will use the fact that $k^5 - k$ is divisible by 5 for any positive integer k (this was proved in **PROBLEM 6.1.2** in the text). First note that, if 5 does not divide k, then 5 must divide $k^4 - 1$. Write $n = 4c + r$, with $r = 0, 1, 2$ or 3, in the expression $1^n + 2^n + 3^n + 4^n$. Then we have

$$1^n + 2^n + 3^n + 4^n$$

$$= 1^{4c+r} + 2^{4c+r} + 3^{4c+r} + 4^{4c+r}$$

$$= 1^r (1^c)^4 + 2^r (2^c)^4 + 3^r (3^c)^4 + 4^r (4^c)^4$$

$$= [1^r ((1^c)^4 - 1) + 2^r ((2^c)^4 - 1) + 3^r ((3^c)^4 - 1) + 4^r ((4^c)^4 - 1)]$$

$$+ (1^r + 2^r + 3^r + 4^r)$$

Since 1^c, 2^c, 3^c, 4^c are not divisible by 5, we must have that $(1^c)^4 - 1$, $(2^c)^4 - 1$, $(3^c)^4 - 1$, $(4^c)^4 - 1$ are all divisible by 5. Thus the term between brackets above is divisible by 5. This implies that $1^n + 2^n + 3^n + 4^n$ is divisible by 5 if and only if $1^r + 2^r + 3^r + 4^r$ is divisible by 5. Recall that $r = 0, 1, 2$ or 3. We have:

If $r = 0$, $1^r + 2^r + 3^r + 4^r = 4$, which is not divisible by 5.

If $r = 1$, $1^r + 2^r + 3^r + 4^r = 10$, which is divisible by 5.

If $r = 2$, $1^r + 2^r + 3^r + 4^r = 30$, which is divisible by 5.

If $r = 3$, $1^r + 2^r + 3^r + 4^r = 100$, which is divisible by 5.

Thus, $1^n + 2^n + 3^n + 4^n$ is divisible by 5 if and only if $n = 4 \cdot c + r$ with $r = 1, 2$ or 3, i.e. if and only if n is not divisible by 4.

6.11. $\ln n$ is just the area under the curve $y = 1/x$ and between $x = 1$ and $x = n$. The sum

$$1 + \frac{1}{2} + \frac{1}{3} + \cdots + \frac{1}{n}$$

is the area of the rectangles under the curve in the figure below.

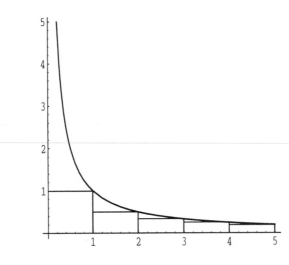

An approximation to

$$1 + \frac{1}{2} + \frac{1}{3} + \cdots + \frac{1}{n} - \ln n$$

is 1 minus the sum of the areas of the regions between each rectangle and the curve. This region is almost a triangle. The kth triangle (i.e. the one over the rectangle of height $1/k+1$) has area

$$\frac{1}{2}\left(\frac{1}{k}-\frac{1}{k+1}\right).$$

Adding up from 1 to n we obtain a telescopic sum that equals

$$\frac{1}{2}\left(1-\frac{1}{n+1}\right).$$

Therefore, the value of

$$1+\frac{1}{2}+\frac{1}{3}+\cdots+\frac{1}{n}-\ln n$$

is approximately

$$1-\left[\frac{1}{2}\left(1-\frac{1}{n+1}\right)\right]=\frac{1}{2}+\frac{1}{2(n+1)},$$

which is definitely less than 4.

A more geometric way to estimate the sum of the areas of the regions between each rectangle and the curve is sliding all these regions horizontally to the first square (the one that has $(0,0)$ as a vertex). Then they will all be disjoint regions contained in a square of area 1, so the sum of their areas must be less than 1.

6.12. One can divide into cases ($x<-1$, $-1\le x<0$, $0\le x<1$, $1\le x<2$, $2\le x$) and then solve the equation in each of the intervals, keeping in mind the restriction on x. The equation is only satisfied for all $x\ge 2$.

6.13. Let S_n be the partial sum up to n in the harmonic series. Then, as we saw in Exercise 11, $S_n\approx\ln n$. The idea is that the sum of the numbers in S_n that have a 7 in the denominator is roughly of size $\ln(n/10)$. Thus, S_n minus the numbers in S_n that have a 7 in the denominator is approximately of size

$$\ln n-\ln(n/10)=\ln\left(\frac{n}{n/10}\right)=\ln 10<\infty.$$

We leave the details to the interested reader.

6.14. Since $n - a_1 - a_2 - \cdots - a_\ell \geq a_{\ell+1}$, we can write

$$\frac{n!}{(a_1)!(a_2)! \cdots (a_k)!}$$

as

$$\frac{n!}{(a_1)!(n - a_1)!} \cdot \frac{(n - a_1)!}{(a_2)!(n - a_1 - a_2)!} \cdots \frac{(n - a_1 - \cdots - a_{k-1})!}{(a_k)!},$$

which equals

$$\binom{n}{a_1} \cdot \binom{n - a_1}{a_2} \cdots \binom{n - a_1 - \cdots - a_{k-1}}{a_k} \cdot (n - a_1 - \cdots - a_{k-1} - a_k)!,$$

which is an integer.

6.15. Note that $p^2 - 1 = (p + 1)(p - 1)$. Since p is prime, it is not divisible by 3, so either $(p + 1)$ or $(p - 1)$ is divisible by 3. In particular, $(p + 1)(p - 1)$ is divisible by 3. Notice that p is not divisible by 2 either, so both $(p + 1)$ and $(p - 1)$ are divisible by 2. In particular, $(p + 1)(p - 1)$ is divisible by 4. Thus, 12 divides $p^2 - 1$. In other words, the remainder when we divide p^2 by 12 is always 1.

6.16. Let p, q, r be the highest powers of 2 that divide x, y, z respectively. Assume, without loss of generality, that $p \leq q \leq r$. The highest power of 2 that can be factored out from the left side of the equation $x^2 + y^2 + z^2 = 2xyz$ is $2p$, and from the right hand side is $p + q + r + 1$. Therefore we must have

$$2p = p + q + r + 1,$$

or

$$p = q + r + 1,$$

which implies $p > q$ and $p > r$, contradicting $p \leq q \leq r$. Therefore the equation has no solutions.

6.17. The solution is 8. One way to solve it is to write 2^{43} as $(1,000 + 24)^4 \cdot 8$ and then multiply.

6.18. If at some point we have some initial amount A at an annual interest rate r, compounded n times a year, then after M years we will have

$$A \cdot \left(1 + \frac{r}{n}\right)^{M \cdot n}.$$

(Note that every time that the interest is compounded we are just multiplying by $(1 + r/n)$.) The child deposits the money for 11 years. In the first case she will have

$$A \cdot \left(1 + \frac{0.05}{365}\right)^{11 \cdot 365} = 1.733A,$$

and in the second case

$$A \cdot \left(1 + \frac{0.051}{52}\right)^{11 \cdot 52} = 1.751A,$$

so the second one is better. That is, it is better to deposit the money at 5.1% interest compounded weekly.

(We are counting 365 days and 52 weeks in a year.)

6.19. One way to proceed is as follows:

$$
\begin{aligned}
S_k &= a + a \cdot r + a \cdot r^2 + \cdots + a \cdot r^k \\
r \cdot S_k &= a \cdot r + a \cdot r^2 + \cdots + a \cdot r^k + a \cdot r^{k+1}.
\end{aligned}
$$

Subtracting the above equations, we obtain

$$S_k - r \cdot S_k = a - a \cdot r^{k+1}.$$

Solving for S_k we finally obtain

$$S_k = \frac{a(1 - r^{k+1})}{1 - r}.$$

6.20. Let A be the amount borrowed, not counting the down payment. Also assume that the first payment is due one month after the money is loaned, and that the interest is compounded monthly at an APR (Annual Percentage Rate) of r percent. Let P be the amount of each monthly payment. After the first payment, the money owed will be

$$A \cdot \left(1 + \frac{r}{12}\right) - P.$$

After the second month it will be

$$\left(A \cdot \left(1 + \frac{r}{12}\right) - P\right) \cdot \left(1 + \frac{r}{12}\right) - P$$

$$= A \cdot \left(1 + \frac{r}{12}\right)^2 - P \cdot \left[\left(1 + \frac{r}{12}\right) + 1\right]$$

It can be proved by induction that, after m months, the amount owed will be

$$A \cdot \left(1 + \frac{r}{12}\right)^m - P \cdot \left[\left(1 + \frac{r}{12}\right)^{m-1} + \cdots + \left(1 + \frac{r}{12}\right) + 1\right].$$

Using the formula of the previous exercise and simplifying, this can be written as

$$A \cdot \left(1 + \frac{r}{12}\right)^m - P \cdot \frac{((1 + r/12)^m - 1)}{r/12}.$$

If we want to pay the loan off in M months, then the last quantity must be 0 for $m = M$, and we obtain the equation

$$A \cdot \left(1 + \frac{r}{12}\right)^M - P \cdot \frac{((1 + r/12)^M - 1)}{r/12} = 0,$$

or

$$P = A \cdot \frac{\frac{r}{12}\left(1 + \frac{r}{12}\right)^M}{\left(1 + \frac{r}{12}\right)^M - 1}.$$

Substituting in the values set in the exercise, we find that she will have to pay \$494.52 a month.

6.21. The area of a sphere of radius r is $4\pi r^2$, and the volume is $4/3\pi r^3$. Thus we have the equation

$$4\pi r = \frac{4}{3}\pi r^3.$$

Solving for r we obtain $r = 3$, which implies that the volume/area of the sphere is 36π.

6.22. Since $x + y + z = 1$ we must have

$$
\begin{aligned}
(x + y + z)^2 &= x^2 + y^2 + z^2 + 2xy + 2yz + 2xz \\
&= 1
\end{aligned}
$$

Since $x^2 + y^2 + z^2$ is a positive number, we must have

$$2xy + 2yz + 2xz < 1,$$

or

$$xy + yz + xz < \frac{1}{2}.$$

6.23. After raising both $10^{1/10}$ and $2^{1/3}$ to the 30th power, we obtain 10^3 and 2^{10}, so it is equivalent to compare these last numbers. Now, $2^{10} = 1,028 > 10^3$, so $2^{1/3}$ is greater than $10^{1/10}$.

6.24. Let $y = 2^x$. Then the equation can be rewritten as

$$8\left(y^2 + \frac{1}{y^2}\right) - 54\left(y + \frac{1}{y}\right) + 101 = 0.$$

Now let $z = y + 1/y$. Note that

$$
\begin{aligned}
z^2 &= \left(y + \frac{1}{y}\right)^2 \\
&= y^2 + 2 + \frac{1}{y^2},
\end{aligned}
$$

so we have $y^2 + \dfrac{1}{y^2} = z^2 - 2$. Thus we can write the equation as

$$8(z^2 - 2) - 54z + 101 = 0,$$

or

$$8z^2 - 54z + 85 = 0.$$

This is a quadratic equation whose solutions are 17/4 and 5/2. Now, since $z = y + 1/y$, we have (solving for y) that

$$y = \frac{z \pm \sqrt{z^2 - 4}}{2}.$$

Thus, the solutions for y are $4, 2, 1/2$ and $1/4$. Since $y = 2^x$, we have that the solutions of the original equation are $-2, -1, 1$ and 2.

6.25. Multiply both sides of the equation by 2. We have

$$
\begin{aligned}
0 &= 2a^2 + 2b^2 + 2c^2 + 2d^2 - 2ab - 2bc - 2cd - 2da \\
&= (a^2 - 2ab + b^2) + (b^2 - 2bc + c^2) + (c^2 - 2cd + d^2) \\
&\qquad\qquad\qquad\qquad\qquad\qquad\qquad + (d^2 - 2da + a^2) \\
&= (a - b)^2 + (b - c)^2 + (c - d)^2 + (d - a)^2
\end{aligned}
$$

Thus, since the last expression is zero only when the summands are zero, we must have $a = b = c = d$.

6.26. If v is the number of fields that each cow grazes bare in one day, then m cows will graze bare mkv fields in k days. In other words, $n = mkv$, or $v = n/mk$. Since v is a constant, we must have

$$\frac{n}{mk} = \frac{n'}{m'k'} = \frac{n''}{m''k''}.$$

6.27. Let a and b be, in order, the last two digits of some number n. Then $n = 100T + 10a + b$, where T is an integer. Then we have

$$n^2 = 100L + 2ab10 + b^2,$$

where L is some integer. Thus, the last two digits of n^2 are the last two digits of $2ab10 + b^2$. If we assume that all the digits of n^2 are 1's, then we must have that the last two digits of $2ab10 + b^2$ are 1's. This implies that $b = 1$ or $b = 9$. If $b = 1$, then the second digit will be the last digit of $2a$, which is even, so it cannot be 1. If $b = 9$, then the second digit will be the last digit of $2a+8$ (8 comes from what we carry over to the second digit when we multiply $9 \cdot 9$), which is also even, so it cannot be 1. Thus no perfect square has 1's as the last two digits of its decimal expression.

6.28. Write

$$N_k = \underbrace{4444\cdots4}_{k\,\text{digits}}\underbrace{888\cdots89}_{k\,\text{digits}}.$$

Then $N_k - 1$ is divisible by

$$\underbrace{1111\cdots1}_{k\,\text{digits}},$$

and we have

$$\frac{N_k - 1}{\underbrace{1111\cdots1}_{k\,\text{digits}}} = 4\underbrace{000\cdots08}_{k\,\text{digits}}.$$

Since $40/6$ gives remainder 4, we can carry on long division to obtain

$$4\underbrace{000\cdots08}_{k\,\text{digits}}/6 = \underbrace{666\cdots68}_{k\,\text{digits}}.$$

Thus we have

$$\begin{aligned}
N_k - 1 &= \underbrace{666\cdots66}_{k\,\text{digits}}\cdot\underbrace{666\cdots68}_{k\,\text{digits}}\\
&= (R_k - 1)(R_k + 1)\\
&= R_k^2 - 1,
\end{aligned}$$

where

$$R_k = \underbrace{666\cdots67}_{k\,\text{digits}}.$$

Therefore we must have $N_k = R_k^2$.

6.29. Take A to be the set of products of all pairs of distinct primes, this is

$$A = \{p \cdot q, \text{where } p \text{ and } q \text{ are distinct primes}\}.$$

Then, if S is any set of primes, then A will contain all products of pairs of elements of S, and the complement of A will contain all the products of triples of elements of S. (Note that the exercise says 'at least two elements of S'.)

6.30. In a set with k elements we have $\binom{k}{m}$ subsets of m elements.

Therefore, the total number of subsets is

$$\binom{k}{0} + \binom{k}{1} + \cdots + \binom{k}{k}.$$

The sum above is exactly the expression of the binomial theorem for $(1+1)^k = 2^k$, so we have 2^k subsets.

6.31. Yes it can. Let us assume the contrary. We know that $\sqrt{2}$ is irrational. By our assumption, we must have that

$$\sqrt{2}^{\sqrt{2}}$$

is irrational. But then

$$\left(\sqrt{2}^{\sqrt{2}}\right)^{\sqrt{2}} = (\sqrt{2})^{\sqrt{2}\cdot\sqrt{2}}$$
$$= (\sqrt{2})^2$$
$$= 2$$

must be irrational, which is false, so we arrive at a contradiction.

6.32. Applying repeatedly the formula for the cosine of the sum of two angles, we find that

$$\cos 5\theta = 16\cos^5\theta - 20\cos^3\theta + 5\cos\theta.$$

Using this equation to solve $\cos(5 \cdot 18°) = 0$, we find that

$$\cos 18° = \sqrt{\frac{(5+\sqrt{5})}{2^3}}.$$

Thus,

$$\cos 36° = 2\cos^2 18° - 1$$
$$= 2\frac{(5+\sqrt{5})}{2^3} - 1$$

$$= \frac{(1 + \sqrt{5})}{4},$$

and

$$\cos 72° = 2\cos^2 36° - 1$$

$$= 2\left(\frac{(5 + \sqrt{5})}{2^3}\right)^2 - 1$$

$$= \frac{(-1 + \sqrt{5})}{4}.$$

This implies

$$\cos 36° - \cos 72° = \frac{\sqrt{5}}{2}.$$

(This solution suggests that there must be a much neater way to find the same result. We encourage the reader to find it.)

6.33. We have $\sin 2\theta = 2\sin\theta\cos\theta = a$. Note that

$$(\sin\theta + \cos\theta)^2 = \sin^2\theta + \cos^2\theta + 2\sin\theta\cos\theta$$

$$= 1 + a.$$

Thus, since θ is acute, $\sin\theta + \cos\theta \geq 0$, and

$$\sin\theta + \cos\theta = \sqrt{1 + a}.$$

6.34. Dividing both sides of the equation $\sin x + \cos x = 1/5$ by $\cos x$ we obtain $\tan x + 1 = \sec x/5$. Using $\tan^2 x + 1 = \sec^2 x$ we find the equation

$$(\tan x + 1)^2 = \frac{\sec^2 x}{25} = \frac{\tan^2 x + 1}{25}.$$

The solution of this equation is either $-\frac{4}{3}$ or $-\frac{3}{4}$. But since $\sin x + \cos x = 1/5 > 0$, we must have $\pi/2 \leq x \leq 3\pi/4$. This implies $\tan x \leq -1$, so the only possible answer is $\tan x = -4/3$.

6.35. Since we are assuming that

$$2 = x^{\left(x^{\left(x^{\cdot^{\cdot^{\cdot}}}\right)}\right)},$$

we have that

$$x^2 = x^{x^{(x^{(x^{.^{.^{.}}}}}}$$
$$= 2$$

This gives $x = \sqrt{2}$.

6.36. Using the same method as in the previous exercise we find that, if

$$x^{(x^{(x^{.^{.^{.}}}}} = a$$

has a solution, then the solution is $x = a^{\frac{1}{a}}$. But we do not know if, for $x = a^{\frac{1}{a}}$, the expression

$$x^{(x^{(x^{.^{.^{.}}}}}$$

actually converges to any number or not. Note that $x^{(x^{(x^{.^{.^{.}}}}}$ is the limit of the sequence

$$a_1 = x$$
$$a_2 = a_1^{a_1}$$
$$a_3 = a_1^{a_2}$$
$$\vdots$$
$$a_k = a_1^{a_{k-1}}$$
$$\vdots$$

We will show that if $a > 1$, this sequence converges (for $a = 1$ it also converges, but this is trivial).

First we will show that this sequence is strictly increasing. Let us do it by induction. First, since $a > 1$, $1 < x = a^{\frac{1}{a}}$. This implies that $a_1 = x < x^x = a_2$. Now assume (induction hypothesis) that $a^{k-1} < a^k$. We want to show that $a^k < a_{k+1}$. But this is clear, since $a^k = x^{a^{k-1}} < x^{a^k} = a^{k+1}$ (here we use again the fact that $x > 1$).

Then we will prove that this sequence is bounded above. By induction again, first note that $a_1 = x = a^{\frac{1}{a}} < a$, so a_1 is bounded above. Now, assume (induction hypothesis) that $a_k < a$. We want to show that $a_{k+1} < a$. But this is clear, since $a^{k+1} = x^{a_k} < x^a = a$.

Now, any strictly increasing sequence that is bounded above has a limit. Thus, for any $a \geq 1$, the number $x = a^{\frac{1}{a}}$ is the only solution of the equation

$$x^{\left(x^{\left(x^{\cdot^{\cdot^{\cdot}}}\right)}\right)} = a.$$

A similar argument shows that this is also true for $0 < a < 1$. The details are left to the reader.

6.37. Let L denote the position of the long hand measured in minutes with $0 \leq L < 60$, and let l denote the position of the short hand measured in hours with $0 \leq l < 12$, (i.e. at three o'clock, $L=0$ and $l = 3$). Then we must have that l and L satisfy the equation

$$k + \frac{L}{60} = l,$$

where $k = 0, 1, 2, \ldots, 11$. The hands meet when $5l = L$. Thus we have that, in this case, l satisfies

$$k + \frac{l}{12} = l.$$

Thus the hands will meet at $l = \dfrac{12k}{11}$ hours and $L = \dfrac{60k}{11}$ minutes, with $k = 0, 1, 2, \ldots, 11$. Now, since for $k = 11$ we obtain $l = 12$ and $L = 60$ (which actually corresponds to $l = 0$, $L = 0$, which is the case $k = 0$), we have that the hands meet 11 times in a 12 hour period.

6.38. The discriminant of the equation is $p^2 - 2q$. We want to show that this number is never a perfect square when p and q are odd integers. Seeking a contradiction, assume that $p^2 - 2q = n^2$ for some integer n. Note that n must be odd. Reordering we obtain $p^2 - n^2 = (p+n)(p-n) = 2q$. Both $p + n$ and $p - n$ are divisible by 2, so $2q$ must be divisible by 4, which implies that q is even, a contradiction.

6.39. Note that $a^3 - b^3 = (a - b)(a^2 + ab + b^2)$. Since a and b are odd, $(a^2 + ab + b^2)$ is odd, and therefore is not divisible by 2^n for any $n \geq 1$. Thus 2^n divides $a^3 - b^3$ if and only if it divides $a - b$.

6.40. We must have $\gamma^2 = \alpha^2 + \beta^2$, or $1 = (\alpha/\gamma)^2 + (\beta/\gamma)^2$. Since both α/γ and β/γ are less than 1 then, for any $n > 2$,

$$\left(\frac{\alpha}{\gamma}\right)^2 > \left(\frac{\alpha}{\gamma}\right)^n \text{ and } \left(\frac{\beta}{\gamma}\right)^2 > \left(\frac{\beta}{\gamma}\right)^n.$$

Thus

$$1 = \left(\frac{\alpha}{\gamma}\right)^2 + \left(\frac{\beta}{\gamma}\right)^2 > \left(\frac{\alpha}{\gamma}\right)^n + \left(\frac{\beta}{\gamma}\right)^n,$$

which implies $\gamma^n > \alpha^n + \beta^n$.

6.41. Assume the contrary. If n is even then the left hand side is even and the right hand side is odd, so n must be odd. If n is odd then, writing the equation as $(n + 2)^3 - n^3 = (n + 1)^3$ we must have, after expanding,

$$8 + 12n + 6n^2 = (n + 1)^3.$$

The right hand side is divisible by 8. The left hand side is not, since that would imply that $6n + 3n^2$ is divisible by 4, which would imply n even. Therefore the equality cannot hold.

6.42. The argument is essentially the same as in Exercise 6.3. Let

$$M = 1 + \frac{1}{3} + \frac{1}{5} + \cdots + \frac{1}{2n + 1},$$

with $n \geq 1$. Let k be the greatest integer such that $3^k \leq 2n + 1$. Note that $k \geq 1$. By the definition of k, we must have

$$3^k \leq 2n + 1 < 3^{k+1}.$$

Now note that none of the numbers $3^k, 3^k + 1, 3^k + 2, \ldots, 2n + 1$ is divisible by 3^k. This is because if one of these numbers is divisible by 3^k, then it must be of the form $\ell \cdot 3^k$, with $\ell \geq 3$ (note that

ℓ must be odd). But then we would have $3 \cdot 3^k \leq n$, contrary to the choice of k. Thus, only one of the denominators in the right hand side of

$$M = 1 + \frac{1}{3} + \frac{1}{5} + \cdots + \frac{1}{2n + 1}$$

(namely 3^k) is divisible by 3^k. Now, let D be the least common multiple of $1, 3, \ldots 2n + 1$. Note that D is divisible by 3^k, but not divisible by 3^{k+1}. Then, writing the expression above with a common denominator, we have

$$
\begin{aligned}
M &= 1 + \frac{1}{3} + \frac{1}{5} + \cdots + \frac{1}{2n+1} \\
&= \frac{D + D/3 + D/5 + \cdots + D/3^k + \cdots + D/(2n+1)}{D}.
\end{aligned}
$$

Finally, note that every summand in the numerator of the last expression is divisible by 3 except for $D/3^k$, which is not. Thus the numerator cannot be divisible by 3 (since all but one of the numbers in the sum are divisible by 3), whereas the denominator is divisible by 3 (since D is divisible by 3^k). But a number that is *not* divisible by 3 divided by a number that *is* divisible by 3 is never an integer. So M cannot be an integer.

6.43. For n even we have that, for some integer k,

$$3^n + 1 = (4 - 1)^n + 1 = 4k + (-1)^n + 1 = 4k + 2,$$

which cannot be divisible by 4, and therefore cannot be divisible by 2^n for $n > 1$.

For n odd, writing $n = 2\ell + 1$, we have that, for some integer k',

$$3^n + 1 = 3 \cdot (8 + 1)^\ell + 1 = 8k' + 3 \cdot 1^\ell + 1 = 8k' + 4,$$

which cannot be divisible by 8, and therefore cannot be divisible by 2^n (note that we are assuming that $n > 1$ and n odd, so $n \geq 3$).

6.44. For each of the five slots we have 3 choices, so the total is $3^5 = 243$. To find how many numbers there are in which each of 1,2,3 occurs at least once, let us subtract from the previous result the number of numbers in which each of 1,2,3 does not occur.

Let A_1 be the set of numbers in which 1 does not occur. The set A_1 has 2^5 elements, since we have five slots and only two choices (2 and 3). Defining A_2 as the set of numbers in which 2 does not occur, and A_3 as the set of numbers in which 3 does not occur, we have that both A_2 and A_3 have 2^5 elements. But note that the number 11111 is in both A_2 and A_3, the number 22222 is in both A_1 and A_3, and the number 33333 is in both A_1 and A_2. Thus, the number of numbers in which each of 1,2,3 does not occur is $3 \cdot 2^5 - 3$ (we subtract 1 for each number that we counted twice). This gives $243 - (3 \cdot 2^5 - 3) = 150$ numbers with the property that each of 1,2,3 occurs at least once.

6.45. Since $5 = 8 - 3$ we can write

$$
\begin{aligned}
5^n + 2 \cdot 3^{n-1} + 1 &= (8-3)^n + 2 \cdot 3^{n-1} + 1 \\
&= 8 \cdot k + (-3)^n + 2 \cdot 3^{n-1} + 1.
\end{aligned}
$$

For n even, $n = 2\ell$,

$$
\begin{aligned}
8 \cdot k + (-3)^n + 2 \cdot 3^{n-1} + 1 &= 8 \cdot k + 5 \cdot 3^{n-1} + 1 \\
&= 8 \cdot k + (8-3) \cdot 3^{n-1} + 1 \\
&= 8 \cdot k' - 3^n + 1 \\
&= 8 \cdot k' - (8+1)^\ell + 1 \\
&= 8 \cdot k'' - 1^\ell + 1 \\
&= 8 \cdot k''',
\end{aligned}
$$

where k, k', k'', k''' are integers.
For n odd, $n = 2j + 1$,

$$
\begin{aligned}
8 \cdot m + (-3)^n + 2 \cdot 3^{n-1} + 1 &= 8 \cdot m - 3 \cdot 3^{n-1} + 2 \cdot 3^{n-1} + 1 \\
&= 8 \cdot m - 3^{n-1} + 1 \\
&= 8 \cdot m - 3^{2j} + 1 \\
&= 8 \cdot m - (8+1)^j + 1 \\
&= 8 \cdot m' - 1^j + 1 \\
&= 8 \cdot m''
\end{aligned}
$$

where m, m', m'' are integers.

6.46. We want to prove that $(1 \cdot 2 \cdots \cdot n)^2 > n^n$. This is equivalent to proving

$$\frac{(1 \cdot 2 \cdots \cdot n)^2}{n^n} > 1,$$

which can be written as

$$\frac{1 \cdot n}{n} \cdot \frac{2 \cdot (n-1)}{n} \cdots \cdots \frac{n \cdot (1)}{n} > 1.$$

Therefore, if we show that each of the factors is greater than or equal to 1 then we are done, as soon as one of the inequalities is strict. Thus we only need to check that

$$k(n - k + 1) \geq n, \quad 1 \leq k \leq n,$$

with strict inequality for at least one k. But this is equivalent to $n \cdot (k-1) - k \cdot (k-1) \geq 0$, which is clearly true, and the inequality is strict for $k = 2$ (recall that $n > 2$).

6.47. Without loss of generality we can assume that a has only two digits, i.e. $a = 10 \cdot b + c$, with $0 \leq b, c \leq 9$. Then $a^2 = 100b^2 + 10(2bc) + c^2$. So we have the conditions:

the ones digit of $2bc$ + the tens digit of $c^2 = 7$.

First this implies that the tens digit of c^2 is odd, which restricts our choices to $c = 4$ or 6. If $c = 4$, then we must have that the ones digit of $2bc$ is 6, or, substituting $c = 4$, that the ones digit of $8b$ is 6, which implies $b = 2, 7$ or 9. If $c = 6$, then we must have that the ones digit of $2bc$ is 4 or, substituting $c = 6$, that the ones digit of $12b$ is 4, which implies $b = 2$ or 7. Thus, the possible values of a are 24, 74, 94, 26, 76, and the ones digits must be 4 or 6.

6.48. $\dfrac{1}{2} + \dfrac{1}{3} + \dfrac{1}{6} = 1$. So, the answer is 2,3,6.

6.49. Expanding the left hand side of the equation

$$(x^2 + ax + b)(x^2 + cx + d) = x^4 + 2x^2 + 2x + 2,$$

we obtain the relations

$$
\begin{aligned}
a + c &= 2 \\
b + ac + d &= 2 \\
ad + bc &= 2 \\
bd &= 2
\end{aligned}
$$

From the last equation we can assume $|b| = 2$, $|d| = 1$. From the first equation we have that a and c have the same parity (i.e. they are both odd or both even). If they are both even, then the left hand side of the second equation is odd, whereas the right hand side is even. Thus they must both be odd. But then the left hand side of the third equation is odd, whereas the right hand side is even. Therefore a, b, c, d cannot all be integers.

6.50. The sum of the reciprocals of all the positive integers that are not divisible by any prime greater than 3 is

$$\left(1 + \frac{1}{2} + \frac{1}{4} + \cdots\right) + \frac{1}{3}\left(1 + \frac{1}{2} + \frac{1}{4} + \cdots\right) + \frac{1}{9}\left(1 + \frac{1}{2} + \frac{1}{4} + \cdots\right) + \cdots$$

Therefore, we can bound the expression in the exercise as follows:

$$
\begin{aligned}
\frac{1}{a_1} + \frac{1}{a_2} + \cdots + \frac{1}{a_n} &< \left(1 + \frac{1}{2} + \frac{1}{4} + \cdots\right) + \frac{1}{3}\left(1 + \frac{1}{2} + \frac{1}{4} + \cdots\right) \\
&\quad + \frac{1}{9}\left(1 + \frac{1}{2} + \frac{1}{4} + \cdots\right) + \cdots \\
&= \left(1 + \frac{1}{2} + \frac{1}{4} + \cdots\right) \cdot \left(1 + \frac{1}{3} + \frac{1}{9} + \cdots\right) \\
&= 2 \cdot \frac{3}{2} \\
&= 3.
\end{aligned}
$$

6.51. Assume $r \neq -1$ (this case is trivial). If $|a| > 1$, then $a \neq 1$ divides $a_{a+1} = a + ar$. If $a = -1, 0$ or 1, then $a + 3r \neq 1$ divides $a_{a+3r+3} = a + ar + 3r^2 + 3r = (a + 3r)(1 + r)$.

6.52. Suppose that $n \cdot (n+1) \cdot (n+2) \cdot (n+3)$ is a perfect square. Then let k be such that $k^2 = n \cdot (n+1) \cdot (n+2) \cdot (n+3)$. Since $n \cdot (n+3) = n^2 + 3n < n^2 + 3n + 2 = (n+1) \cdot (n+2)$, we have

$$(n \cdot (n+3))^2 < n \cdot (n+1) \cdot (n+2) \cdot (n+3) < (n+1) \cdot (n+2),$$

which implies

$$n \cdot (n+3) = n^2 + 3n < k < n^2 + 3n + 2 = (n+1) \cdot (n+2)$$

Thus, we must have $k = n^2 + 3n + 1$. But then it is impossible that $k^2 = n \cdot (n+1) \cdot (n+2) \cdot (n+3)$, since n divides $n \cdot (n+1) \cdot (n+2) \cdot (n+3)$ but it does not divide k^2.

6.53. There are $\binom{5}{4} \cdot 4! = 5 \cdot 24$ numbers of this kind (we have $\binom{5}{4}$ ways to choose the 4 digits and $4!$ permutations of each choice). Write the numbers in a column as follows:

$$
\begin{array}{c}
1234 \\
1235 \\
1243 \\
2543 \\
\vdots
\end{array}
$$

Then observe that each of $1, 2, 3, 4, 5$ appears $1/5$ of the time in the units slot, $1/5$ of the time in the tens slot, $1/5$ of the time in the hundreds slot, etc. Thus, the sum of all these numbers is

$$24 \cdot 1111 + 24 \cdot 2222 + 24 \cdot 3333 + 24 \cdot 4444 + 24 \cdot 5555 = 399,960.$$

6.54. For r a real number, let (r) denote its fractional part. For example, $(15.324) = 0.324$. Consider the set of numbers

$$\{(a), (2a), (3a), \ldots, ((n-1)a)\}.$$

This is a set of $n-1$ numbers in the interval $[0, 1)$. If one of these numbers, say $(k \cdot a)$, is less than or equal to $1/n$, that means that $k \cdot a$ differs from some integer by at most $1/n$. Similarly, if one of these numbers, say $j \cdot a$, is greater than or equal to $(n-1)/n$, that also means that $k \cdot a$ differs from some integer by at most $1/n$. So let us assume that all the numbers $\{(a), (2a), (3a), \ldots, ((n-1)a)\}$ lie in the interval $(1/n, (n-1)/n)$. The length of this interval is $(n-2)/n$. Since the set $\{(a), (2a), (3a), \ldots, ((n-1)a)\}$ contains $n-1$ numbers, two of them must differ by at most $1/n$ (this is just the pigeonhole principle: we have $(n-1)$ numbers and $(n-2)$ sub-intervals of length $1/n$ each, so two of the numbers must lie in the same sub-interval). Denote these two numbers by $(s \cdot a)$ and $(t \cdot a)$, and assume that $s < t$. Then the number $(t-s)a$ is also in the set $\{(a), (2a), (3a), \ldots, ((n-1)a)\}$ and it differs from some integer by at most $1/n$. We leave it to the reader to convince herself/himself of this fact.

6.55. $17x + 17y - (9x + 5y) = 8x + 12y = 4(2x + 3y)$. Thus $2x + 3y$ is divisible by 17 if and only if $9x + 5y$ is divisible by 17.

6.56. We see that $2^n + 1 = (3 - 1)^n + 1 = 3K + (-1)^n + 1$ for some integer K. Thus $2^n + 1$ is divisible by 3 if and only if $(-1)^n + 1$ is zero, i.e. if and only if n is odd.

6.57. The least value n can have is $p_1 \cdot p_2 \cdots p_k$, where p_i are consecutive primes in increasing order with $p_1 = 2$. Thus $p_i \geq 2$, which implies $n \geq 2^k$, which is equivalent to $\log n \geq k \log 2$.

6.58. First, since the numbers $a_1 \cdot a_2, a_2 \cdot a_3, \ldots, a_n \cdot a_1$ are either 1 or -1, and their sum is 0, we must have as many 1's as -1's in the list, so n must be even. Write $n = 2k$. On the other hand, since k of the numbers $a_1 \cdot a_2, a_2 \cdot a_3, \ldots, a_n \cdot a_1$ are -1, and the rest are 1, we have

$$(a_1 \cdot a_2)(a_2 \cdot a_3) \ldots (a_n \cdot a_1) = (-1)^k.$$

But we also have

$$(a_1 \cdot a_2)(a_2 \cdot a_3) \ldots (a_n \cdot a_1) = a_1^2 \cdot a_2^2 \cdots \cdots a_n^2.$$

Since each of the numbers a_i is either 1 or -1, its square is 1, so the left hand side of the last equation equals 1. This implies

$$1 = (-1)^k.$$

Therefore, $k = 2\ell$ for some positive integer ℓ, which implies $n = 4 \cdot \ell$. Thus n is divisible by 4.

6.59. The number $n^{n-1} - 1$ can be written as $((n-1)+1)^{n-1} - 1$. Using the binomial theorem, we obtain

$$
\begin{aligned}
n^{n-1} - 1 &= ((n-1)+1)^{n-1} - 1 \\
&= \binom{n-1}{0}(n-1)^{n-1} + \binom{n-1}{1}(n-1)^{n-2} + \cdots \\
&\quad + \binom{n-1}{n-3}(n-1)^1 + \binom{n-1}{n-2} + 1 - 1 \\
&= \binom{n-1}{0}(n-1)^{n-1} + \binom{n-1}{1}(n-1)^{n-2} + \cdots \\
&\quad + \binom{n-1}{n-3}(n-1)^2 + (n-1)(n-1),
\end{aligned}
$$

where we have used the fact that $\binom{n-1}{n-2} = n - 1$. Finally, note that every term of the last sum is divisible by $(n-1)^2$. Thus, $(n-1)^2$ divides $n^{n-1} - 1$.

6.60. The correct analog is

$$
\cos \frac{\alpha}{2} \cos \frac{\alpha}{4} \cos \frac{\alpha}{8} \cos \frac{\alpha}{16} = \frac{\sin \alpha}{16 \sin (\alpha/16)}.
$$

The proof, using the result in Problem 6.3.5, reduces to

$$
\frac{\sin \alpha}{8 \sin (\alpha/8)} \cos \frac{\alpha}{16} = \frac{\sin \alpha}{16 \sin (\alpha/16)}.
$$

Multiplying both sides by $16 \sin (a/16)$, we obtain

$$
\frac{\sin \alpha}{8 \sin (\alpha/8)} 16 \cos \frac{\alpha}{16} \sin \frac{\alpha}{16} = \sin \alpha.
$$

Using the formula for the sine of the double angle, we get

$$
\frac{\sin \alpha}{8 \sin (\alpha/8)} 8 \sin (\alpha/8) = \sin \alpha,
$$

which is a tautology. Since the process can be reversed, the equality is proved.

6.61. Note that $n^2(n^2 - 1)(n^2 - 4) = (n - 2)(n - 1)n^2(n + 1)(n + 2)$ (five consecutive integers). One of these numbers is divisible by 4, another one by 2, two of them by 3 and one of them by 5. Thus, the product is divisible by $4 \cdot 2 \cdot 3 \cdot 3 \cdot 5 = 360$.

6.62. See *Principles of Mathematical Analysis,* by Walter Rudin or *Real Analysis and Foundations,* by Steven G. Krantz.

6.63. We can write $a = 12r$ and $b = 12s$, with r and s relatively prime. Then $12rs = 432$, or $rs = 36 = 2 \cdot 2 \cdot 3 \cdot 3 \cdot$. Since r and s do not have common factors, we must have $r = 9$, $s = 4$ (or vice versa). Thus $a = 108$, $b = 48$ (or vice versa).

6.64. Denote the number by t. Note that $9 \cdot t + 1 = 10^{92}$. So we have

$$t = \frac{(10^{41} - 1)(10^{41} + 1)}{9}.$$

Thus t is factorable.

6.65. Expanding $(m + n + k)^3$, we have

$$
\begin{aligned}
(m + n + k)^3 &= k^3 + 3k^2m + 3km^2 + m^3 + 3k^2n + 6kmn \\
&\quad + 3m^2n + 3kn^2 + 3mn^2 + n^3 \\
&= m^3 + n^3 + k^3 + 3K,
\end{aligned}
$$

where K is some integer. Thus $(m + n + k)^3$ is divisible by 3 if and only if $m^3 + n^3 + k^3$ is.

6.66. Suppose that $a^2 - b^2 = p^2$. Then we have $(a + b)(a - b) = p^2$, which is a factorization of p. But since p is a prime, we must have $a + b = p$ and $a - b = 1$. Solving for a and b, we obtain

$$\left(\frac{p + 1}{2}\right)^2 - \left(\frac{p - 1}{2}\right)^2 = p.$$

Note that, by construction, this is the only possible way to write p as the difference of two squares.

6.67. Write $m^2 = p^2 - n^2 = (p+n)(p-n)$. Call $a = (p+n)$, $b = (p-n)$. Note that for each value of a and b we can find p and n by just solving the equation that defines a and b. This actually gives

$$p = \frac{a+b}{2}$$
$$n = \frac{a-b}{2}$$

Since p and n are integers, our only condition on a and b is that they be either both odd or both even (so that both $a+b$ and $a-b$ are divisible by 2).

Thus, for each m, and for any two integers a, b with same parity satisfying

$$m^2 = a \cdot b,$$

we can find two integers p, n in a unique way so that the equation

$$m^2 + n^2 = p^2$$

is satisfied. This classifies all the Pythagorean triples.

6.68. Four iterations give an accuracy of 1 decimal place. Five iterations give an accuracy of 2 decimal places.

6.69. The factorization with real coefficients is

$$x^8 + x^4 + 1 = (1 - x + x^2)(1 + x + x^2)(x^2 - x\sqrt{3} + 1)(x^2 + x\sqrt{3} + 1).$$

6.70. Using the formula $\cos^2 x = (1 + \cos 2x)/2$ we obtain

$$\begin{aligned}
\cos^4 x &= \left(\frac{1 + \cos 2x}{2}\right)^2 \\
&= \frac{1}{4}\left(1 + 2\cos 2x + \cos^2 2x\right) \\
&= \frac{1}{4}\left(1 + 2\cos 2x + \frac{1 + \cos 4x}{2}\right) \\
&= \frac{1}{8}\left(3 + 4\cos 2x + \cos 4x\right)
\end{aligned}$$

6.71. We have to find the sum of the sum of the sum of the digits of 4444^{4444}. We will use the fact that, if a number can be written as $9c + r$, then the sum of its digits can be written as $9c' + r$ (same r). First note that $4444 = 9 \cdot 494 - 2$. Thus

$$4444^{4444} = (9 \cdot 494 - 2)^{4444} = 9 \cdot \ell + 2^{4444},$$

where ℓ is an integer. Now, since $2^3 = 8 = 9 - 1$, and $4444 = 3 \cdot 1481 + 1$, we have

$$
\begin{aligned}
2^{4444} &= 2^{3 \cdot 1481 + 1} \\
&= 8^{1481} \cdot 2 \\
&= (9 - 1)^{1481} \cdot 2 \\
&= (9 \cdot k - 1) \cdot 2 \\
&= 9 \cdot k' - 2,
\end{aligned}
$$

where k, k' are integers. Therefore, the number 4444^{4444} can be written as

$$9 \cdot \ell + 9 \cdot k' - 2 = 9 \cdot (\ell + k') - 9 + 7 = 9 \cdot k'' + 7,$$

where k'' is an integer.

Thus, we must have that the sum of the sum of the sum of the digits of 4444^{4444} is also of the form $9c + 7$. The number 4444^{4444} is less than 10000^{4444}, which has 4445 digits. Thus, the sum of the digits of 4444^{4444} is less than or equal to $9 \cdot 4445 = 40005$. Now, 40005 has 5 digits, so the sum of its digits is at most $5 \cdot 9 = 45$. Finally, the sum of the digits of a number that is less than 45 is always a one-digit number. Therefore, we have that the sum of the sum of the sum of the digits of 4444^{4444} is a one-digit number of the form $9c + 7$. The only possibility is that this number is 7 itself. Thus, the sum of the sum of the sum of the digits of 4444^{4444} is 7.

6.72. First note that the quantity

$$\frac{1 \cdot 3 \cdots (2n - 1) \cdot (2n + 1)}{2 \cdot 4 \cdots 2n \cdot (2n + 2)}$$

is less than or equal to

$$\frac{2 \cdot 4 \cdots 2n \cdot (2n+2)}{3 \cdot 5 \cdots (2n+1) \cdot (2n+3)}.$$

This is because $1/2 < 2/3$, $3/4 < 4/5$, $\ldots$, $(2n+1)/(2n+2) < (2n+2)/(2n+3)$. Thus, if we multiply these two quantities, we will obtain something that is greater than the square of the first quantity. This is

$$\left(\frac{1 \cdot 3 \cdots (2n-1) \cdot (2n+1)}{2 \cdot 4 \cdots 2n \cdot (2n+2)}\right)^2$$

$$< \left(\frac{1 \cdot 3 \cdots (2n-1) \cdot (2n+1)}{2 \cdot 4 \cdots 2n \cdot (2n+2)}\right) \cdot \left(\frac{2 \cdot 4 \cdots 2n \cdot (2n+2)}{3 \cdot 5 \cdots (2n+1) \cdot (2n+3)}\right)$$

But the right hand side simplifies to $1/(2n+3)$. Thus we have

$$\frac{1 \cdot 3 \cdots (2n-1) \cdot (2n+1)}{2 \cdot 4 \cdots 2n \cdot (2n+2)} < \frac{1}{\sqrt{2n+3}} < \frac{1}{\sqrt{n}},$$

which is what we wanted to prove.

6.74. From $m + n = m \cdot n$ we have that m divides n and vice-versa. Thus $m = n$. This implies $2m = m^2$, or $m = 0$ or 2. Thus the only possible solutions are $m = n = 0$ and $m = n = 2$.

6.75. We will show that $n^{11} - n$ is divisible by 11, the case $n^{13} - n$ being similar. First, since every number n can be written as $11 \cdot m + q$, $0 \le q < 11$, we have, using the binomial theorem,

$$\begin{aligned} n^{11} - n &= (11 \cdot m + q)^{11} - (11 \cdot m + q) \\ &= 11 \cdot A + q^{11} - q. \end{aligned}$$

Thus $n^{11} - n$ is divisible by 11 if and only if $q^{11} - q$ is divisible by 11 for $0 \le q < 11$. This is easy to check by hand.

In general, it is true that $n^k - n$ is divisible by k for any integers n and k. This is called 'Fermat's Little Theorem' (not to be confused with 'Fermat's Last Theorem'). The proof of this result is not too difficult using modular arithmetic.

Chapter 7

A Miscellany

7.1. Here is one example

$$4 + 5 + 9 + 13 + \frac{72}{8} + 60.$$

We cannot do this without using fractions. We explain the reason in the following way:

$$0 + 1 + 2 + \cdots + 9 = 45.$$

Notice that 45 is divisible by 9. We know that any number is divisible by 9 if and only if the sum of its digits is divisible by 9. In fact, any positive integer, divided by 9, has the same remainder as the remainder of the sum of it is digits divided by 9. This means that if we subtract the sum of the digits of a positive integer from the number itself, the result is going to be divisible by 9. For example, $28 - (2 + 8) = 18$, $49 - (4 + 9) = 36$, and so on. Suppose N is the sum of a collection of positive integers in which all the digits $0, 1, \ldots, 9$ have been used and they have been used just once. Then $N - (0 + 1 + \cdots + 9) = N - 45$ should be divisible by 9. So N cannot be 100 because $100 - 45 = 55$, and 55 is not divisible by 9.

7.2. Let us refer to the two men by M and m and to their wives by A, B and a, b respectively. We also assume that they want to cross the river from side 1 to side 2. Here is a possible solution:

(a) First a and b go to the side 2.

(b) Then a takes the boat back to side 1.

(c) A and B go to side 2.

(d) b takes the boat to side 1.

(e) a and b go to side 2.

(f) A takes the boat to side 1.

(g) M and m go to side 2.

(h) B takes the boat to side 1.

(i) Finally A and B go to side 2.

Now they all have crossed the river, and the problem is solved.

7.3. Let us use the same conventions as in the previous exercise. It is obvious that the solution for the previous problem still works for this one, but it does not use the entire capacity of the boat. One possible solution which uses the entire capacity of the boat is as follows:

(a) A, M and m go to side 2.

(b) A takes the boat back to side 1.

(c) A, a and b go to side 2.

(d) A takes the boat to side 1 (if you think that A is doing all the work you can replace her with any of the other women).

(e) Finally A and B go to side 2.

7.4. We refer to the three men by m_1, m_2, m_3 and to their wives respectively by a_1, b_1, c_1, a_2, b_2, c_2 and a_3, b_3, c_3. The two sides of the river are side 1 and side 2. Here is one possible solution:

(a) First a_1, b_1 and c_1 go to side 2.

(b) Then a_1 takes the boat to side 1.

(c) a_1, a_2 and b_2 go to side 2.

(d) b_2 takes the boat to side 1.

(e) b_2, c_2 and a_3 go to side 2.

(f) a_3 takes the boat to side 1.

(g) m_1, m_2 and m_3 go to side 2.

(h) c_2 takes the boat to side 1.

(i) Now there are four women in side 1 with the boat. Three of them go to side 2. One of them comes back and takes the remaining woman to the other side.

7.5. We refer to people as in the previous exercise and assume that they want to transfer from side 1 to side 2. Also, let us call the boats T_1 and T_2.

(a) a_1 and b_1 in T_1, and a_2 and b_2 in T_2 go to side 2.

(b) a_1 in T_1 and a_2 in T_2 take the boats back to side 1.

(c) a_1 and c_1 in T_1, and a_2 and c_2 in T_2 go to side 2.

(d) b_1 and b_2 take the boats T_1 and T_2 back to side 1.

(e) m_1 and b_1 in T_1 and, m_2 and b_2 in T_2 go to side 2.

(f) Now in side 1, there are only four people, the third man with his three wives, and the boat is on side 2. Since the first two men cannot send any of their wives to side 1, otherwise their wives would be without chaperone in the other side of the river with m_3, they have to take the boats back to side 1 themselves.

(g) In T_1, m_3 and one of his wives go to side 2, and in T_2, m_1 and m_2 go to side 2.

(h) Now any two women in side 2 can take the boats to side 1 and bring the two remaining women to side 2.

7.6. This problem is similar to **PROBLEM** 7.2.3 in the text. We divide the nodes into two groups: Those with an odd number of edges emanating from them and those with an even number of edges. If a node has an odd number of edges emanating from it, then the trace either begins in that node without ending there or ends there without having started at that node. So we can have

at most two nodes with an odd number of edges (in fact there can
be only two nodes of this type or none at all). In Figure 1 there
are far more than two such nodes, so no solution is possible. For
more explanation see **PROBLEM** 7.2.3.

7.7. The solution is: B, C, D, G, I, J, L, M, N, O, P, Q, R, S, U, V, W,
Z. For further explanation see **PROBLEM** 7.2.3.

7.8. Suppose that X and Y are positive numbers and $X + Y = 100$.
Then

$$\begin{aligned} XY &= X(100 - X) \\ &= 100X - X^2 \\ &= 2500 - 2500 + 100X - X^2 \\ &= 2500 - (50 - X)^2. \end{aligned}$$

Since $(50 - X)^2$ is positive or zero, $2500 - (50 - X)^2$ is no greater
than 2500. So XY is no greater than 2500 and as a result, $XY = 3000$ is impossible.

7.9. Obviously if we allow the interiors of the circles to intersect the
sides of the square then the problem is solvable even by using only
one circle. So the assumption is that the interior of a circle does
not intersect the sides of the square. At most the boundary of a
circle can touch one or two of the sides. Obviously if we are to
fill the square by finitely many circles, some of them must touch
the sides. A line and a circle touch at exactly one point. With
finitely many circles we can at most cover finitely many points on
the sides of the square; since a square has infinitely many points
on its sides, all but finitely many points on the sides of the square
will be outside all the circles.

7.10. Similar reasoning as in the previous exercise.

7.11. If an equilateral triangle has side 1, then it has area $\sqrt{3}/4$. Thus if
it could be turned into a square then that square would have side
$\sqrt[4]{3}/2$. Also it would have diagonal $\sqrt[4]{3} \cdot \sqrt{2}/2$. Note that both
this side length and this diagonal length are less than 1. So, if the

triangle could be cut with a single cut so that the two pieces can form a square, then all three sides of the triangle (having length 1) would have to be cut (to achieve the desired shortness). This is clearly impossible with a single cut.

7.12. Suppose Amelia starts walking from right to left and Debbie starts walking from left to right. We assume that the length of the road is λ. The first time Amelia and Debbie pass each other, Amelia has walked 720 feet and Debbie $\lambda - 720$ feet. The second time they meet (both on their way back), Amelia has covered a total distance of $\lambda + 400$ while Debbie has covered a total distance of $2\lambda - 400$. Since their speed has not changed, the two relevant ratios $720/(\lambda - 720)$ and $(\lambda + 400)/(2\lambda - 400)$ are equal. Setting them equal and solving for λ we obtain $\lambda = 1760$.

7.13. We start with 1. We count what we see: ONE 1 or "1" 1. We add these to the sequence, and we get

$$1, 1, 1.$$

We count again. Now we see "THREE" 1's. We add 3, 1 to the sequence, to obtain

$$1, 1, 1, 3, 1.$$

Now we see "FOUR" 1's and "ONE" 3. By adding these to the sequence we get

$$1, 1, 1, 3, 1, 4, 1, 1, 3.$$

Now there are "SIX" 1's, "TWO" 3's and "ONE" 4. The sequence now grows to

$$1, 1, 1, 3, 1, 4, 1, 1, 3, 6, 1, 2, 3, 1, 4.$$

Continuing this way we get

$$1, 1, 1, 3, 4, 1, 1, 3, 6, 1, 2, 3, 1, 4, 8, 1, 3, 3, 2, 4, 1, 6, 1, 2.$$

So "?" stands for 1.

7.14. Let us assume imaginary stations at the end of each day's journey and call them station 1, station 2 and etc. We also change the problem a little: Suppose the explorer wants to walk as far as station n; assuming he can take three days' provisions at the frontier of the desert as many times as it may be necessary, what is the minimum amount of provisions he needs to go as far as station n?

We will not solve the above in complete generality, but only for particular examples. If $n = 1$, the explorer obviously will need two days' food. For $n = 2$, since he cannot take four days' provisions all at once, he takes as much as he can (three days' food) and walks to station 1, stores one day's food there and comes back to the origin. Again, he takes another load and goes to station 1. Now he has enough food to last him till station 2 and his return journey to the frontier of the desert. He has fetched food twice, and that is six days' worth of provisions.

If he wants to go as far as station 3, then by repeating the previous case, we can conclude that he needs at least six days' food at station 1. In addition, he needs enough food for one more day for his return journey. Thus, in total, he needs seven days' food at station 1. To get this amount, he has to fetch food 6 times at the frontier, and that is 18 days' worth of provisions.

In the same way, we can calculate that in order to get to station 4, he would need 54 days' worth of provisions.

In our problem, he only has enough food for 30 days'. Obviously, he cannot get to station 4. So he has to stop somewhere between station 3 and 4. We leave it to the reader to show that he can go half way between the two stations, or in other words 28 miles deep into the desert.

7.16. A cube has 12 edges. We can think of them as twelve different places where we can place the sticks. So the question is: in how many different ways can we place these sticks? First we have twelve choices; after choosing the first stick, we have 11 to choose from, and then we have 10 and so on. So the number of different combinations is $12 \times 11 \times 10 \cdots \times 1$. The above number is called

12 factorial and is denoted by 12! (k! $= k \times k - 1 \times \cdots \times 2 \times 1$).
Moreover, once a position is chosen for a stick, it can be placed
in two different ways (imagine that the sticks are marked on one
end). Taking this into account we get $2^{12} \cdot 12!$. On the other
hand, two ways of placing the sticks could result in the same
cube (one could be the rotation of the other). So we have to make
adjustments for this. To do so, we divide the number above, by
the number of different ways one cube can occupy the same space.
We can imagine that we have a cube which has sides of different
colors. We can also agree which position is to be called the front,
back, up, down, right and left. For each side there are 4 different
ways in which that side can be in the front. We have 6 sides, so
in total there are $4 \times 6 = 24$ different ways that the same cube
can occupy the same space. Hence, we get

$$2^{12}\frac{12!}{24}.$$

7.17. In general, $(a^3+b^3)/2$ is not equal to $((a + b)/2)^3$ because $Y \to Y^3$
is not a linear function.

7.19. Suppose 1000 people decide to spin the wheel. According to the
odds approximately 800 are going to win \$800. Hence, 200 of
them will win nothing. So, on average, each person wins 640
dollars. So on average, people get more money by spinning the
wheel. So it seems that it is wiser to spin the wheel.

7.20. We expect that, if the sample is random, the ratio of the marked
bees in the sample to the size of the sample is roughly the same as
the ratio of all marked bees to the population of bees. Assuming
this, we can estimate that the population of the beehive must be
around $100(100/6) \approx 1666$.

One fault with this method is that it actually counts the number
of bees present in the beehive, rather than the total population
(some bees could be out collecting pollen). And at the same time,
there is no justification for the assumption that the part of the
population of the beehive which is not in the beehive has the
same proportion of marked bees as the part which is inside the

beehive. Thus, even the estimate we get for the number of bees in the beehive may not be quite accurate.

Also it does not take into account particular habits of bees. At mating time or feeding time the population could segregate in special ways. One should do this experiment several times.

7.22. a) If 2 has a rational square root then it has a positive rational square root. Call that number μ and write $\mu = m/n$, where the fraction is in least terms (all common factors divided out).

b) This is clear.

c) We know that 2 divides m^2. On the other hand, if a prime number divides the product of two numbers, it must divide at least one of them. In our case 2 has to divide one of m and m, which is to say it has to divide m.

d) This is clear.

e) This is the same as part c).

f) This is clear.

7.23. Suppose there is such a rational number. Let us refer to it by r. So the assumption is that $r^2 = 8$. If r is a rational number then so is $r/2$. But

$$\left(\frac{r}{2}\right)^2 = \frac{r^2}{4} = \frac{8}{4} = 2.$$

So we just found a rational number whose square is 2. But this contradicts Exercise 7.22, in which we have shown that there is no such rational number. Hence, we conclude that the assumption that $\sqrt{8}$ is rational is incorrect.

7.24. Suppose that k is an integer and $k = (m/n)^2$. If k is not a perfect square (of an integer) then at least one of its prime factors occurs to an odd power. We have $kn^2 = m^2$. All prime factors on the right occur to even powers. But some prime factor on the left occurs to an odd power. Contradiction!

7.25. Let r and s be two different rational numbers, r being the smaller one. Then $a = (s - r)/\sqrt{2}$ is an irrational number, because otherwise $\sqrt{2} = (s - r)/a$ implies that $\sqrt{2}$ is a rational number which contradicts Exercise 7.22. On the other hand, a is positive and is less than $s - r$. So $r + a$ is greater than r and less than s. At the same time, $r + a$ is an irrational number, otherwise $a = r + a - r$ would be rational, but we just showed that this is not the case.

7.26. Let r and s be two different irrational numbers, r being the smaller one. If $a = s - r$ is greater than 1, then there is an integer between r and s. An integer is certainly a rational number. On the other hand if their difference is less than 1, no matter how small this difference is, there is always an integer n, such that $n(s - r) > 1$. So there is an integer m between nr and ns. Hence, m/n is between r and s.

7.28. There are certain circles of latitude (near the south pole) that have length 1 mi., 1/2 mi., 1/3 mi., 1/4 mi., etc. Number these circles C_1, C_2, C_3, C_4, etc. Now suppose the explorer begins one mile north of circle C_j. She walks south 1 mile, so ends up on the circle C_j. Then she walks east 1 mile, which means she goes exactly j times around the circle. Then she walks north one mile and returns to her starting place.

Chapter 8

Real Life

8.1. Let R be the radius of the circular arc in the problem, and let ϕ be the angle that it spans in radians. Then we must have

$$2R\sin\frac{\phi}{2} = 5280$$

and

$$R\phi = 5281.$$

We want to find

$$h = R - R\cos\frac{\phi}{2}.$$

Using the first equation we have

$$h = R - \sqrt{R^2 - (2640)^2}.$$

Thus, everything reduces to finding R. This is no easy task, but it can be done. The first two equations give

$$R\sin\frac{2640.5}{R} = 2640.$$

One can find an approximate numerical solution using a computer package. Another possibility is to use a computer to plot, for example, the graph of $R\sin 2640.5/R - 2640$, and then patiently finding the interval in which the function is approximately 0. Using this method and the computer package Mathematica, we found $R \approx 78335.08051$. This gives $h \approx 44.4985$ feet, which is much larger than one would expect.

8.2. First, let us find a formula for $i(N)$. Only one number between 0 and 9 begins with a 1. Therefore, $i(N) = 1$ if $0 \leq N \leq 9$. Let us find $i(N)$ for N between 10 and 99. We have $i(10) = 2, i(11) = 3, \ldots, i(19) = 11$. After 19 there are no numbers between 10 and 99 that start with a 1. Thus $i(N) = 11$ for N between 20 and 99.

Now let us find $i(N)$ for N between 100 and 999. First note that, if $100 \leq N \leq 999$, we have

$$i(N) = \#\{\text{numbers that begin with a 1 in } [0, 90]\}$$
$$+ \#\{\text{numbers that begin with a 1 in } [100, N]\}$$
$$= \quad i(99) + \#\{\text{numbers that begin with a 1 in } [100, N]\}$$

Now, all the numbers that begin with a 1 between 100 and N are in the interval $[100, 200]$. Also, the first number in this interval that begins with a 1 is 100, and the last one is 199, that makes the 100th number that begins with a 1 in this interval. This can be expressed as follows:

$$\#\{\text{numbers that begin with a 1 in } [100, N]\}$$
$$= \begin{cases} N - 99 & \text{if } 100 \leq N \leq 199 \\ 100 & \text{if } 200 \leq N \leq 999. \end{cases}$$

Thus we have

$$i(N) = \begin{cases} 11 + N - 99 & \text{if } 100 \leq N \leq 199 \\ 111 & \text{if } 200 \leq N \leq 999. \end{cases}$$

Following the same procedure as above, we can find the following general formula:

$$i(N) = \begin{cases} \dfrac{10^k - 1}{9} + N - 10^k + 1 & \text{if } 10^k \leq N \leq 2 \cdot 10^k - 1 \\ \dfrac{10^{k+1} - 1}{9} & \text{if } 2 \cdot 10^k \leq N \leq 10^{k+1} - 1. \end{cases}$$

Note that, in each of these intervals, the minimum ratio between $i(N)$ and N is achieved when N is largest, i.e. when $N = 10^{k+1} - 1$. In this case we have

$$\frac{i(N)}{N} = \frac{(10^{k+1} - 1)/9}{10^{k+1} - 1} = \frac{1}{9}.$$

Thus we have that $i(N)/N \geq 1/9$ for all N.

Similarly, in each of these intervals, the maximum ratio between $i(N)$ and N is achieved when $N = 2 \cdot 10^k - 1$. In this case we have

$$\frac{i(N)}{N} = \frac{1}{2 \cdot 10^k - 1} \left(\frac{10^k - 1}{9} + 2 \cdot 10^k - 1 - 10^k + 1 \right) \approx 0.56.$$

Thus $i(N)/N$ fluctuates approximately between $1/9$ and 0.56.

Now notice that the quantity $i(N)/N$ gives the proportion of numbers that start with 1 between 1 and N, i.e. the probability that, when we choose a number at random between 1 and N, its first digit will be a 1. We want to find this probability independent of N. To this end, we have to find the average of all these probabilities. This is, we need to calculate

$$\lim_{M \to \infty} S_M,$$

where

$$S_M = \frac{1}{M} \sum_{N=1}^{M} \frac{i(N)}{N}.$$

First, it is necessary to show that such a limit exists. This can be proved as follows: Let us find the limit as $N \to \infty$ of the difference of two members of the sequence:

$$S_{M+K} - S_M = \frac{1}{M+K} \sum_{N=1}^{M+K} \frac{i(N)}{N} - \frac{1}{M} \sum_{N=1}^{M} \frac{i(N)}{N}$$

$$= \frac{1}{M(M+K)} \left(M \cdot \sum_{N=1}^{M+K} \frac{i(N)}{N} - (M+K) \cdot \sum_{N=1}^{M} \frac{i(N)}{N} \right)$$

$$= \frac{1}{M(M+K)} \left(M \cdot \sum_{N=M+1}^{M+K} \frac{i(N)}{N} - K \cdot \sum_{N=1}^{M} \frac{i(N)}{N} \right)$$

$$= \frac{1}{M(M+K)} \left(M \cdot \sum_{N=M+1}^{M+K} \frac{i(N)}{N} \right) - \frac{1}{M(M+K)} \left(K \cdot \sum_{N=1}^{M} \frac{i(N)}{N} \right)$$

Since $1/9 \leq i(N)/N \leq 0.56$, the first term in the last expression can be bounded as follows:

$$\frac{1}{M(M+K)}\left(M\cdot\sum_{N=M+1}^{M+K}\frac{1}{9}\right) \leq \frac{1}{M(M+K)}\left(M\cdot\sum_{N=M+1}^{M+K}\frac{i(N)}{N}\right)$$

$$\leq \frac{1}{M(M+K)}\left(M\cdot\sum_{N=M+1}^{M+K}0.56\right).$$

This gives

$$\frac{1}{M(M+K)}\left(\frac{MK}{9}\right) \leq \frac{1}{M(M+K)}\left(M\cdot\sum_{N=M+1}^{M+K}\frac{i(N)}{N}\right)$$

$$\leq \frac{1}{M(M+K)}\left(0.56\cdot MK\right).$$

When $M \to \infty$, both sides of this inequality tend to 0, so the term in the middle must also tend to 0.

Similarly, the second term can be bounded as follows:

$$\frac{1}{M(M+K)}\left(K\cdot\sum_{N=1}^{M}\frac{1}{9}\right) \leq \frac{1}{M(M+K)}\left(K\cdot\sum_{N=1}^{M}\frac{i(N)}{N}\right)$$

$$\leq \frac{1}{M(M+K)}\left(K\cdot\sum_{N=1}^{M}0.56\right).$$

Thus we have

$$\frac{1}{M(M+K)}\left(\frac{MK}{9}\right) \leq \frac{1}{M(M+K)}\left(K\cdot\sum_{N=1}^{M}\frac{i(N)}{N}\right)$$

$$\leq \frac{1}{M(M+K)}\left(0.56\cdot MK\right).$$

Again, when $M \to \infty$, both sides of the last inequality tend to 0, so the term in between must also tend to 0.

Thus we have proved that, for any value of K

$$\lim_{M\to\infty}(S_{M+K} - S_M) = 0.$$

This implies that the sequence S_M converges (the meaning of $\lim_{M \to \infty}(S_{M+K} - S_M) = 0$ is that the difference between two terms of the sequence tends to 0 as M tends to infinity, so there must be a point where the sequence accumulates).

Finally, now that we have proved that the limit exist, we have to find it. It is extremely hard to find the limit exactly, but it is not so hard to give a quite accurate approximation. First, since we know that the limit does exist, it is enough to find the limit of a subsequence. Let us find

$$\lim_{k \to \infty} S_{10^{k+1}-1}.$$

In the process, we will use the approximation

$$\frac{1}{1} + \frac{1}{2} + \frac{1}{3} + \cdots + \frac{1}{N} = \ln(N+1) + E(N),$$

where $E(N)$ is an error term between 0 and 1 (this is a very similar approximation to that of Exercise 11 in Chapter 6).

Note that this implies that, for $M > N$,

$$
\begin{aligned}
\sum_{\ell=N}^{M-1} \frac{1}{\ell} &= \ln M - \ln N + (E(M) - E(N)) \\
&= \ln\left(\frac{M}{N}\right) + E',
\end{aligned}
$$

where $E' = (E(M) - E(N))$ is an error term between 0 and 1 (note that $E(N)$ increases with N—see the figure in the solution of Exercise 11 of Chapter 6).

Using the formula for $i(N)$ above, it is easy to find a general formula for $i(N)/N$:

$$
\frac{i(N)}{N} = \begin{cases} 1 - \dfrac{8 \cdot (10^k - 1)}{9N} & \text{if } 10^k \leq N \leq 2 \cdot 10^k - 1 \\[2mm] \dfrac{10^{k+1} - 1}{9N} & \text{if } 2 \cdot 10^k \leq N \leq 10^{k+1} - 1. \end{cases}
$$

Then we have

$$\sum_{N=1}^{10^{k+1}-1} \frac{i(N)}{N}$$

$$= \sum_{\ell=0}^{k} \sum_{N=10^\ell}^{10^{\ell+1}-1} \frac{i(N)}{N}$$

$$= \sum_{\ell=0}^{k} \left(\sum_{N=10^\ell}^{2\cdot 10^\ell-1} \frac{i(N)}{N} + \sum_{N=2\cdot 10^\ell}^{10^{\ell+1}-1} \frac{i(N)}{N} \right)$$

$$= \sum_{\ell=0}^{k} \left[\sum_{N=10^\ell}^{2\cdot 10^\ell-1} \left(1 - \frac{8\cdot(10^\ell-1)}{9N} \right) + \sum_{N=2\cdot 10^\ell}^{10^{\ell+1}-1} \frac{10^{\ell+1}-1}{9N} \right]$$

$$= \sum_{\ell=0}^{k} \left[10^\ell - \frac{8\cdot(10^\ell-1)}{9} \cdot \sum_{N=10^\ell}^{2\cdot 10^\ell-1} \frac{1}{N} + \frac{10^{\ell+1}-1}{9} \cdot \sum_{N=2\cdot 10^\ell}^{10^{\ell+1}-1} \frac{1}{N} \right]$$

Using the logarithmic approximation explained above, we obtain

$$\sum_{\ell=0}^{k} \left[10^\ell - \frac{8\cdot(10^\ell-1)}{9} \cdot \left(\ln\left(\frac{2\cdot 10^\ell}{10^\ell}\right) + E' \right) \right.$$
$$\left. + \frac{10^{\ell+1}-1}{9} \cdot \left(\ln\left(\frac{10^{\ell+1}}{2\cdot 10^\ell}\right) + E'' \right) \right]$$

$$= \sum_{\ell=0}^{k} \left[10^\ell - \frac{8\cdot(10^\ell-1)}{9} \cdot (\ln 2 + E') + \frac{10^{\ell+1}-1}{9} \cdot (\ln 5 + E'') \right]$$

$$= \frac{10^{k+1}-1}{10-1} - \frac{8(\ln 2 + E')}{9} \cdot \frac{10^{k+1}-1}{10-1} + \frac{8(k+1)(\ln 2 + E')}{9}$$
$$+ \frac{(\ln 5 + E'')}{9} \cdot \frac{10^{k+2}-10}{10-1} - \frac{(k+1)(\ln 5 + E'')}{9}$$

To find $\lim_{k\to\infty} S_{10^{k+1}-1}$ we have to divide the last quantity by $10^{k+1}-1$ and then take the limit as $k \to \infty$. Now, since

$$\lim_{k\to\infty} \frac{1}{10^{k+1}-1} \cdot \frac{8(k+1)(\ln 2 + E')}{9} = 0$$

and

$$\lim_{k\to\infty} \frac{1}{10^{k+1}-1} \cdot \frac{(k+1)(\ln 5 + E'')}{9} = 0,$$

it only remains to find (after simplifying),

$$\lim_{k\to\infty} \frac{1}{10^{k+1}-1} \left(\frac{10^{k+1}-1}{9} - \frac{8(10^{k+1}-1)(\ln 2 + E')}{81} \right.$$
$$\left. + \frac{(10^{k+2}-10)(\ln 5 + E'')}{81} \right).$$

This limit equals

$$\frac{1}{9} - \frac{8 \cdot (\ln 2 + E')}{81} + \frac{10 \cdot (\ln 5 + E'')}{81},$$

or, reordering.

$$\frac{1}{9} + \frac{10\ln 5 - 8\ln 2}{81} + \frac{10E'' - 8E'}{81},$$

which is, approximately, $0.241348 + E'''$, where E''' is an error term roughly between -0.0987654 and 0.123457. Thus, given a number, the average probability that its first digit is a 1 is at least $0.142583 \approx 1/7$.

8.4. See the discussion of this problem in [PAUL1].

8.5. The idea is that figures change depending on what year we take as a base for the price of the commodities. See also the nice exposition in the book *How to Lie with Statistics*, by Darrell Huff.

The cost of living has gone up: Take last year as the base year. This means that, to find a percentage of how much prices have changed, we take the price of the commodities last year to be 100. That means that the price of bread is 200% higher now than it was last year, and the price of milk had a drop of 50%. The average of 200 and 50 is 125. Therefore, the cost of living has gone up 25%.

The cost of living has gone down: Take this year as the base year. Assume that the price of the commodities this year is 100. Last year, milk used to cost 200% as much as it does now, and bread was selling for 50% as much. The average is 125. Thus, the cost of living was 25% higher last year.

The cost of living did not change: Take last year as the base year, but use geometric average instead of arithmetic. The price of milk now is 50% of last year's, and the price of bread is 200%. The geometric average is $\sqrt{50 \cdot 200} = 100$. Thus, the prices are 100% of last year's prices; in other words, the cost of living has not changed.

8.6. See the beautiful account of this and other related problems in the book *How to Lie with Statistics*, by Darrell Huff, Chapter 8.

8.7. Same reference as in the previous exercise, page 82. It is clear that one is making 1% on total sales (1 cent for each item, one dollar per item). Also, the money invested is 99 cents and the total profit is 365 cents, approximately 365% on money invested.

8.8. A word with four letters has an approximate linear measure of 1 inch. One line of text has approximately 15 words of this length, and one page has approximately 35 lines. This means that for each page we use $1 \cdot 15 \cdot 34 = 480$ inches worth of ink, or 40 feet. one mile is 5280 feet, so with that ball pen we will be able to write $5280/40 = 132$ pages.

8.9. According to the Encyclopædia Britannica, hair grows at a speed of 0.5 inches per month. There are $24 \cdot 30 = 720$ hours in a month, and $12 \cdot 5280 = 63360$ inches in a mile. Therefore, hair grows at a rate of

$$\frac{0.5 \cdot 720}{63360} = 0.315657 \cdot 10^{-4} \text{ miles per hour.}$$

8.10. The total amount paid after n months on a principal P at a rate r, compounded monthly, is

$$P\left(1 + \frac{r}{12}\right)^n.$$

(See Exercise 20 in Chapter 6.)

In our case, this gives 106.168.

8.11. In this problem we use Bayes' formula (see Chapter 3 for details):

$$\mathbf{Pr}\{\text{Irving has VD } | \text{ test } +\}$$
$$= \frac{\mathbf{Pr}\{\text{test } + | \text{ has VD }\} \cdot \mathbf{Pr}\{ \text{ has VD}\}}{\mathbf{Pr}\{+| \text{ VD }\} \cdot \mathbf{Pr}\{\text{VD}\} + \mathbf{Pr}\{+| \text{ no VD }\} \cdot \mathbf{Pr}\{ \text{ no VD}\}}$$
$$= \frac{0.98 \cdot 0.005}{0.98 \cdot 0.005 + 0.02 \cdot 0.995}$$
$$= 0.197581,$$

which is remarkably low. See also the discussion of this problem in [PAUL1, p. 89]

The probability that Irving has VD when one test is negative is

$$\mathbf{Pr}\{\text{Irving has VD } | \text{ test } -\}$$
$$= \frac{\mathbf{Pr}\{\text{test } - | \text{ has VD }\} \cdot \mathbf{Pr}\{ \text{ has VD}\}}{\mathbf{Pr}\{-| \text{ VD }\} \cdot \mathbf{Pr}\{\text{VD}\} + \mathbf{Pr}\{-| \text{ no VD }\} \cdot \mathbf{Pr}\{ \text{ no VD}\}}$$
$$= \frac{0.02 \cdot 0.005}{0.02 \cdot 0.005 + 0.98 \cdot 0.995}$$
$$= 0.000102543.$$

Now, since different tests are uncorrelated, the probability of Irving having VD after two negative tests is the square of the probability of Irving having VD after one negative test, i.e. it is the square of 0.000102543.

Thus, the probability that Irving has VD after two negative tests is approximately 10^{-8}.

8.12. It is the same analysis as in the previous exercise. Now we do not know, though, what is the percentage of people that lie (note that before we were given that 0.5% of the people had VD). Assuming that this percentage gives a probability of p, we have

$$\mathbf{Pr}\{\text{lying } | \text{ test } +\}$$
$$= \frac{\mathbf{Pr}\{\text{test } +| \text{ lying }\} \cdot \mathbf{Pr}\{\text{lying}\}}{\mathbf{Pr}\{+| \text{ lying}\} \mathbf{Pr}\{\text{lying}\} + \mathbf{Pr}\{+| \text{ not lying}\} \mathbf{Pr}\{\text{not lying}\}}$$
$$= \frac{0.75 \cdot p}{0.75 \cdot p + 0.25 \cdot (1 - p)}$$

$$= \frac{1.5p}{0.5 + p}.$$

If, for example, $p = 0.1$, this gives

$$\mathbf{Pr}\{\text{lying} \mid \text{test} +\} = 0.25.$$

8.13. The important point in this exercise is that he takes *the first train* that comes along. This means that if the train to New York always arrives 2 minutes before the train to Philadelphia, he will very rarely take this last train. In fact, unless he arrives in this two-minute interval, he will always end up in New York. There are 3 trains per hour, i.e. 3 two-minute intervals of this kind per hour. Hence the probability that he takes the train to Philadelphia is $2 \cdot 3/60 = 1/10$. In other words, he visits his girlfriend in New York nine times as often as his girlfriend in Philadelphia.

8.14. In one hour, Sam can type 1/10th of the manuscript, and George can type 1/5th. Together, they would type $1/10 + 1/5 = 3/10$th's of the manuscript in one hour. Therefore, together they would type the entire manuscript in 10/3 hours, or 3 hours and 20 minutes.

8.15. He counted the paces to his friend's house. Then, while he was with his friend, he timed the walk by counting up to the same number at approximately the same rhythm as his paces. He noted the time when he left his friend's house and added the time length of the walk when he set up his clock.

8.16. The probability that they have no common acquaintances is

$$\frac{\binom{250 \cdot 10^6}{1500} \cdot \binom{250 \cdot 10^6 - 1500}{1500}}{\binom{250 \cdot 10^6}{1500}^2} \approx 0.99104.$$

(This value was found by simplifying the fraction first and then using a computer. No machine can handle numbers as big as $(250 \cdot 10^6)!$) Thus, the probability that they have a common acquaintance is ≈ 0.00896.

The second probability can be found as follows. We will assume that the probability that two people do not know each other is independent from the fact that they may not know other people. This approximation is valid, given that the sample space (the people in America) is very large. We have then that, since the probability that two people know each other is $1500/(250 \cdot 10^6)$, then the probability that they do not know each other is

$$p = 1 - \frac{1500}{250 \cdot 10^6}.$$

Let us find the probability that the two people A and B are not connected by a chain of acquaintances as in the statement of the exercise. We first assume, of course, that A and B have no common acquaintances (which happens with the probability given above). We have to multiply this probability by the probability that none of the 1,500 acquaintances of A knows any of the 1,500 acquaintances of B, given that we have 3,000 different people (since we assume that A and B have no common acquaintances). This probability can be estimated as follows: we have to multiply all the probabilities that each of the friends of A has of not knowing any of the friends of B. Since there are $1,500 \cdot 1,500 = 2,250,000$ pairings of friends of A with friends of B, this probability is

$$p = 2,250,000 \approx 1.3709 \cdot 10^{-6}.$$

Thus, the probability that they are not connected by a chain of acquaintances as in the statement of the exercise is approximately $(1 - 0.00896) \cdot 1.3709 \cdot 10^{-6} = 1.3586210^{-6}$. This gives that the probability of such a scheme is greater than 99%. See [PAUL1, p. 38] for details.

8.17. The maximum number of hairs in the human scalp is about 500,000. Since there are about 10 million people in New York, by the pigeonhole principle two people must have the same number of hairs. See [PAUL1, p. 42] for a more detailed discussion.

8.18. The probability that no heads come up before the 15th flip is $(1/2)^{14}$, so this is the probability of winning the game. The game will be fair or favorable if

$$p \geq \frac{L}{L+W},$$

where p is the probability of winning, L is the amount you pay if you lose and W is the amount you receive if you win (see **Chapter 3** for details).

Therefore we have to verify the inequality

$$\left(\frac{1}{2}\right)^{14} \geq \frac{10}{1,000,010}.$$

This inequality holds. The game is actually very favorable.

8.19. The probability that it does not rain is $(1/2) \cdot (1/2) = 1/4$. Therefore the probability that it does rain is $1 - 1/4 = 3/4$, or 75%.

8.20. The height of a newborn increases by about 20cm in the first year. Let us assume that this rate is approximately constant for the first year. A year has $365 \cdot 24 \cdot 60 = 525600$ minutes. 20cm are $0.2 \cdot 10^{-3}$km. Thus the rate of baby growth in km/min is

$$\frac{0.2 \cdot 10^{-3}}{525600} = 3.80518 \cdot 10^{-10}.$$

8.21. According to the Encyclopædia Britannica, there are approximately 60 mililitres of blood per kilogram in the human body. Take the average weight of a person to be 50 kilograms (note that children are included). There are 250 million people in the U.S., which will give $60 \cdot 50 \cdot 250 \cdot 10^6 = 75 \cdot 10^{10}$ mililitres, or 750 cubic meters. Now you have to find the radius r in meters of the base circle of Busch Stadium, and then solve for h in the equation

$$\pi r^2 h = 750.$$

8.22. The average number of hairs in an average person ranges between 100,000 and 150,000, according to Encyclopædia Britannica. Let us take 125,000, the average of the two quantities. Measure the length of one representative of your hairs. Let us say it is 3 inches, or 0.25 feet. Then we have that the total length of all the hairs is $125,000 \cdot 0.25 = 31250$ feet, or about 5.9 miles.

8.23. The probability that he drops a hamburger is 0.3. We assume that the events are independent, i.e. for each hamburger the probability is 0.3 no matter what happened with the other hamburgers. Thus the probability that he drops four of the next ten is

$$0.3^4 \cdot 0.7^6 \cdot \binom{10}{4} \approx 0.200121.$$

8.25. Since the maximum number of hairs in the human scalp is about 500,000, each person can have from 0 to 500,000 hairs, so there are 500,001 possibilities. Thus, we need at least 500,002 people to be absolutely sure that two of them have the same number of hairs.

The problem of finding how many people we need in order to find two scalps with the same number of hairs is similar to the birthday problem in the text. We will assume that the distribution of number of hairs in the population is uniform and ranges between 0 and $500,000$, and that the number of hairs of two different people are uncorrelated variables. We have

$$\mathbf{Pr}\{\text{at least two of } N \text{ people have same no. of hairs}\}$$
$$= 1 - \mathbf{Pr}\{N \text{ people have different number of hairs}\}$$
$$= 1 - \frac{500,001 \cdot 500,000 \cdots (500,001 - N + 1)}{500,001^N}$$

With the help of a computer, we found that this value is greater than $1/2$ for $N \geq 833$ and less than $1/2$ for $N < 833$. Thus, we need 833 people in order to have probability at least $1/2$ of finding two people with the same number of hairs.

8.26. The price of postage in 1996 can be calculated to be about 40 cents per letter, including the envelope and the paper for the letter. Let us say that the total expenses are \$1 per letter (this includes possible telephone calls, copies of documents, etc). The total number of letters he sent is $2,000 + 1,000 + 500 + 250 = 3,750$. Hence he has to earn \$103,750 in order to make a profit of \$100,000. Now, $103,750/250 = 415$. Thus he has to charge at least \$415 to each one of the 250 people. For a discussion of this and other related problems see [PAUL1, p.42].

8.28. The total amount paid by the company is $2 \times 20,000 + 20 \times 10,000 + 200 \times 1,000 + 1000 \times 250 = \$690,000$. Dividing by the number of people gives that the company pays \$13.50 per policy. Since they want to clear one million dollars, each person has to pay $1,000,000/50,000 = \$20$ on top of the \$13.50. Thus the premium should be \$33.50.

8.29. The average of both years does not equal the sum of the averages of each year. For example, if Forrest got 4 fish in 1 trip in 1987 and 3 fish in 1 trip in 1988, and if Bubba got 31 fish in 8 trips in 1987 and 5 fish in 2 trips in 1988, then we have

$$\frac{4}{1} > \frac{31}{8}$$
$$\frac{3}{1} > \frac{5}{2} \tag{8.1}$$

But the average for the two years for Forrest is

$$\frac{4+3}{1+1} = \frac{7}{2} = 3.5,$$

and for Bubba,

$$\frac{31+5}{8+2} = \frac{36}{10} = 3.6.$$

We see that Bubba's average over the two years is greater.

See also [PAUL1, p.44] for this and other related problems.

8.30. Let us denote by '>' the sentence 'wins against'. i.e., '$\alpha > \beta$' means 'α wins against β'. Then we have

$$\mathbf{Pr}\{\alpha > \beta\} = \mathbf{Pr}\{\alpha \text{ got a 4}\}$$
$$= \frac{2}{3}$$

$$\mathbf{Pr}\{\beta > \gamma\} = \mathbf{Pr}\{\gamma \text{ got a 2}\}$$
$$= \frac{2}{3}$$

$$\mathbf{Pr}\{\gamma > \delta\ \} \ = \ \mathbf{Pr}\{\gamma \text{ got a 6}\} + \mathbf{Pr}\{\gamma \text{ got a 2 and } \delta \text{ got a 1}\}$$

$$= \ \frac{1}{3} + \frac{2}{3} \cdot \frac{1}{2}$$

$$= \ \frac{2}{3}$$

$$\mathbf{Pr}\{\delta > \alpha\} \ = \ \mathbf{Pr}\{\delta \text{ got a 5}\} + \mathbf{Pr}\{\delta \text{ got a 1 and } \alpha \text{ got a 0}\}$$

$$= \ \frac{1}{2} + \frac{1}{2} \cdot \frac{1}{3}$$

$$= \ \frac{2}{3}$$

8.33. If the rain is falling vertically, then clearly the best strategy is not to move, since in this way we expose less surface area to the rain. If the rain is falling at an angle then the best strategy is to move in the direction of the rain and at the same horizontal speed as the rain. The idea is that if we move with the rain, exactly at the same speed and the same direction, it is essentially as if the rain were falling vertically and we were standing still.

8.34. Let us denote time by t, beginning at noon, and let T_0 be the time elapsed from the beginning of the snow till noon. Note that $T_0 > 0$. Let $x(t)$ denote the area covered by the plow at time t, measured in blocks. Finally, let r be the rate at which the snow falls (in inches per hour) and let R denote the rate at which the plow clears the show (measured in blocks × inches / hours).

The area covered by the plow in a short interval of time $[t, t + \Delta t]$ is $x(t + \Delta t) - x(t)$. The volume of snow cleared in this interval of time is approximately the area that the plow covered times the height of the snow at time t, which is $(t + T_0)r$. On the other hand, the volume of snow cleared in this interval of time equals $R \cdot \Delta t$. Thus we have the equation

$$(x(t + \Delta t) - x(t))(t + T_0)r = R \cdot \Delta t.$$

Dividing both sides by Δt and letting $\Delta t \to 0$, we obtain

$$x'(t)(t + T_0)r = R,$$

or, reordering

$$x'(t) = \frac{R}{(t + T_0)r}.$$

Now, integrating both sides, we obtain

$$x(t) = \frac{R}{r} \ln C(t + T_0),$$

where C is a constant.

To determine the value of C, first use the fact that at noon (i.e. at $t = 0$) the plow had not covered any area, so we have $x(0) = 0$. This gives the equation

$$\frac{R}{r} \ln C(t + T_0) = 0,$$

which implies $C = 1/T_0$. Substituting the value of C and simplifying, we obtain

$$x(t) = \frac{R}{r} \ln \left(\frac{t}{T_0} + 1\right).$$

On the other hand, we have that at $t = 1$, the plow had covered 2 blocks (so $x(1) = 2$), and at $t = 2$, the plow had covered a total of 3 blocks (so $x(2) = 3$). Therefore, we have

$$\frac{R}{r} \ln \left(\frac{1}{T_0} + 1\right) = 2$$
$$\frac{R}{r} \ln \left(\frac{2}{T_0} + 1\right) = 3$$

Dividing the first equation by the second and simplifying (note that R and r cancel), we obtain

$$3 \ln \left(\frac{1}{T_0} + 1\right) = 2 \ln \left(\frac{2}{T_0} + 1\right),$$

or

$$\ln \left(\frac{1}{T_0} + 1\right)^3 = \ln \left(\frac{2}{T_0} + 1\right)^2.$$

Cancelling the logarithms on both sides we get

$$\left(\frac{1}{T_0} + 1\right)^3 = \left(\frac{2}{T_0} + 1\right)^2.$$

Expanding and simplifying we get

$$4T_0 + 4T_0^2 = 1 + 3T_0 + 3T_0^2,$$

which gives the equation

$$T_0^2 + T_0 - 1 = 0.$$

The solutions of this equation are $\dfrac{-1 \pm \sqrt{5}}{2}$. Since we know that it started snowing *before* noon, T_0 must be positive. Thus, we must have

$$T_0 = \frac{-1 + \sqrt{5}}{2} \approx 0.618034.$$

Hence, it started snowing approximately 0.618034 hours before noon, or at 11 and 0.381966 hours, which is approximately 22 minutes and 55 seconds past 11.

8.35. Assuming that the cost of manufacturing a tire is proportional to its durability, if they made tires that last less than 20,000 miles, then they would be giving out a lot of money in replacements of tires. Similarly, if they were making tires that lasted longer than 40,000 miles, then the prices are too low. The number that maximizes profit in these cases is the average; in our case, they should make tires that last 30,000 miles. Note that, in average, only half the people that purchased 30,000 mile-guaranteed tires will return them.

8.37. See [PAUL1, p.28] and [REN] for a discussion of these matters.

8.39. The reason they are round is so that they do not fall through the hold and injure the worker below. Clearly a miniscule perturbation of round will still do the job, but make it more difficult to replace the cover properly into its place when the job is done (these covers are very heavy).

8.40. North has to come first, as in the figure below. Otherwise the freeway would have to double back over the South turnoff, increasing construction costs.

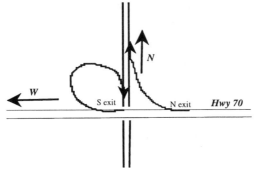

8.44. The probability of having 47 consecutive boys born in a family line is

$$\left(\frac{1}{2}\right)^{47} \approx 7.105428 \cdot 10^{-15}.$$

8.45. The probability of a monkey typing Hamlet, assuming that the typewriter has 35 keys and that Hamlet has 500,000 characters is 1/35 to the power 500,000. Let us call this probability p. If we have a very long chain of characters of length K, Hamlet could appear in the first 500,000 characters, or in characters 2 to 500,001, or in characters 3 to 500,002, etc. Thus, Hamlet could start in the first, second, third, up to the $K - 500,000 + 1$ character. Thus we have $K - 499,999$ ways of obtaining Hamlet in the string of characters. This implies that the probability of finding Hamlet in this string is $p \cdot (K - 499,999)$. Since we want this last probability to be 0.5, we must have

$$K = 499,999 + \frac{1}{2 \cdot p} = 499,999 + \frac{1}{2 \cdot 35^{500,000}} \geq 10^{727272}.$$

Assuming that in a year the monkeys may type, say, a billion characters per year (this is more than 400,000 pages, since each page has about 2,400 characters), the monkeys would take more than 10^{727263} years to type Hamlet by chance. The age of the universe is estimated to be much, much less than this figure. See [PAUL1, p.75] for an interesting and humorous discussion of this and other related problems. We also recommend to the reader the short story *The Library of Babel*, by Jorge Luis Borges, in which ideas similar to the one posed in this problem are discussed in a literary and poetic manner.

8.46. The reason why heads occur more often in the second experiment is that the coin is not a very flat cylinder; instead, the edge has a very slight slant, so the shape of the coin is actually a fulcrum of a cone. This is easy to see with the following experiment: put a penny on the table as in the second experiment. Then use a ruler or something with a 90 degree angle on it, place one side on the table and the other against the tail side of the coin, as in the figure below. One can see that the coin is actually slightly tilted toward the tail side. When we slam the table, the coin loses its equilibrium and falls, but since it is tilted toward the tail side it is more likely to fall with the face side up. The same explains the first experiment: the coin is heavier on the face side (since the radius of this side is greater than the radius of the other). To find an equilibrium while rotating, the coin tilts slightly toward the face side. When it finally stops rotating and falls, it is more likely to fall with the tail side down.

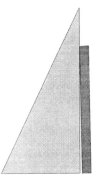

8.48. Let l, j, c and s denote the price in cents of the licorice, the jerky, the chips and the soda respectively. According to the statement of the exercise we must have

$$\frac{l}{100} \cdot \frac{j}{100} \cdot \frac{c}{100} \cdot \frac{s}{100} = \frac{l}{100} + \frac{j}{100} + \frac{c}{100} + \frac{s}{100} = 7.11.$$

We can rewrite the last equations as

$$l + j + c + s = 711$$

and

$$l \cdot j \cdot c \cdot s = 2^6 \cdot 3^2 \cdot 5^6 \cdot 79.$$

Now, 79 must divide l, j, c or s. Assume that 79 divides s. Then we can write $s = 79k_s$. Note that $k_s = 1, 2, 3, 4, 5, 6$ or 8. The equations can now be written as

$$l + j + c = 79(9 - k_s)$$

and

$$l \cdot j \cdot c \cdot k_s = 2^6 \cdot 3^2 \cdot 5^6.$$

Using trial and error or by means of a computer we can find that the four prices are \$3.16, \$1.20, \$1.50 and \$1.25. Note that

$$3.16 + 1.20 + 1.50 + 1.25 = 3.16 \cdot 1.20 \cdot 1.50 \cdot 1.25 = 7.11.$$